The World Food Problem

David Grigg

Second edition

BLACKWELL
Oxford UK & Cambridge USA

Copyright © David Grigg 1985, 1993

The right of David Grigg to be identified as author of this work has been asserted in accordance with the Copyright, Designs and Patents Act 1988.

First published 1985
First published in paperback 1986
Reprinted 1988, 1989
Second edition 1993

Blackwell Publishers
108 Cowley Road
Oxford   OX4 1JF
UK

238 Main Street
Cambridge, Massachusetts 02142
USA

British Library Cataloguing in Publication Data

A CIP catalogue record for this book is available from the British Library.

Library of Congress Cataloging-in-Publication Data

Grigg, David B.
    The world food problem / David Grigg. — 2nd ed.
        p.   cm.
    Includes index.
    ISBN 0-631-17632-2. — ISBN 0-631-17633-0 (pbk.)
    1. Food supply. I. Title.
HD9000.5.G74   1993
363.8—dc20   93-3274
                                                        CIP

Typeset in 10 on 12 pt Plantin
by Graphicraft Typesetters Ltd, Hong Kong
Printed in Great Britain by Page Brothers, Norwich

This book is printed on acid-free paper

The World Food Problem

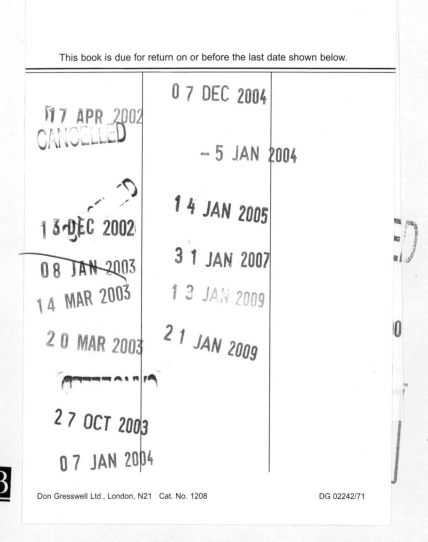

For Jill, Susan, Catherine and Stephen,
with much love

Contents

List of figures

List of tables

Acknowledgements

My interest in the problems discussed in this book was first stimulated by Vic Dennison and was encouraged by Benny Farmer in the 1950s and since, and by the late Charles Fisher in the 1960s. I hope this book may encourage and stimulate others, as their teaching and example did – and does – encourage me.

I am grateful to Mrs Kate Schofield, Mrs Jean Walters, Mrs Helen Doncaster and Mrs Margaret Gray for typing the manuscript with their usual efficiency and great dispatch. I am similarly beholden to Mr Graham Allsopp and Mr Paul Coles for drawing the maps and graphs.

I am grateful for permission to reproduce and redraw illustrations from *Ceres*, the FAO review on agriculture and development (figures 2.5, 8.3 and 11.4); A. and E. Weber and the *European Review of Agricultural Economics* (figures 3.1 and 3.2); HMSO (figure 7.1); George Philip & Son Limited (figure 8.2); Methuen & Co. (figure 8.4); the National Research Council of Canada (figure 8.10); Armand Colin Editeur (figure 9.1); and Holt-Saunders Limited (figure 10.1).

I am very grateful to the University of Sheffield Travel Fund, the Royal Society and particularly the British Academy and the Leverhulme Trust for grants that enabled me to visit the libraries in FAO headquarters in Rome, and also to library staff there for their help.

David Grigg

1

Introduction

For much of human history the majority of mankind has been under-nourished, and for much of human history this, it would seem, has been stoically accepted; there are few records left to tell us. But from the eighteenth century, writers began to argue that hunger could be overcome, sought explanations of its continued existence and proposed solutions to the problem. Hunger and malnutrition were still widespread in Europe in the nineteenth century, but by the 1930s were greatly diminished. But, for the most part, when Europeans discussed this problem they confined their attention to their own continent and the people of European origin settled overseas. Thus when Sir William Crookes predicted in 1899 that the world's wheat supply would soon prove insufficient he was thinking mainly of the food supplies of Europeans, as was Yves Guyot in 1904 and G. B. Roorbach in 1917. Even the League of Nations, which in 1928 declared that two-thirds of the world's population was inadequately fed, in its subsequent three-volume report concerned itself largely with Europe.[1] Indeed it was not until the Second World War that Europeans and Americans began to concern themselves with the world food problem. In a meeting at Hot Springs in Colorado in 1943, a conference looked forward to a world free from hunger.[2]

Thus, since the end of the Second World War there has been much discussion of the world food problem. Indeed it might sometimes seem that a shortage of food and a prevalence of hunger have only existed since 1945. This is obviously not so. It is more that changes since 1945 have brought the problem to the world's attention far more forcibly than before, for a number of reasons.

The first is the foundation in 1945 of the United Nations and its subsidiary, the Food and Agriculture Organization (FAO). The latter was formed to try and increase agricultural output and solve the problem of hunger. Later international organizations such as the World

Bank and voluntary associations like Oxfam have promoted knowledge about the world food problem and attempted practical solutions.

Second, more people have written and spoken about the problem than at any time in the past. The spread of literacy in the developing world, the growth of newspapers and the extension of radio and television have made most people in Europe and North America aware of the problem, and, of course, the remarkable increase in travel has allowed more Europeans to visit Africa, Asia and Latin America. All these changes have also made Africans, Asians and Latin Americans – or at least some of them – aware of their condition in comparison with the West, an awareness heightened by the struggle for and achievement of independence in the period since 1945.

Third, the problem of hunger – the world food problem – has received far more attention from the academic world. Since 1945 universities have multiplied in all parts of the world, and both traditional and new disciplines have turned their attention to the economic problems of the developing world and in particular to the problems of food supplies. In all subjects the growth of knowledge since 1945 has been remarkable, and not least in the study of the world food problem.

Fourth, the causes and solutions of the world food problem have been a matter of ideological controversy, and this alone would have been enough to attract attention. In both the popular view and the opinion of many academic writers, the world food problem since 1945 has been due to the population explosion in the developing countries; many believe that something should be done to restrain population growth. But attempts to introduce family planning methods into Africa, Asia and Latin America met much opposition in the 1950s and 1960s. Catholics and other religious groups were opposed to the use of contraceptives; they were joined, in a curious alliance, by socialists who believed that it was not population growth that caused hunger, but the imperfections of capitalist society.[3] In the 1970s this verbal conflict subsided, but was revived in the 1980s. In the United States small but vocal groups opposed the American government's support of family planning programmes in the developing countries.

A further ideological controversy concerned the role of economic growth in reducing hunger. Most writers would agree that poverty is the prime cause of hunger, but there has been much debate about how poverty in the developing countries should be overcome. In 1950 most of the countries in Afro-Asia and Latin America were primarily agrarian and were poor. It was thought then both by economists and politicians that economic growth required industrialization, but there were sharp differences over the way this could be achieved. Socialist writers and governments believed that the national ownership of resources combined with central planning was the most efficient way of achieving industrialization, economic growth and hence the reduction of poverty and hunger. Others believed that the market economies of Western Europe and the United States were a better model for the developing world. Whether the fall of one-party communist states in Eastern Europe in 1989 will influence the socialist or quasi-socialist states in the developing world remains to be seen.

For these and other reasons the world food problem has become a familiar topic since the end of the Second World War, and there is no end of books and articles upon the subject. Best known are those which have prophesied doom. The idea that population growth will at some time in the future outrun food supplies, and universal starvation arrive, has been much publicized, from W. Vogt's *The Road to Survival*, published in 1949, through *Famine 1975*, by W. and P. Paddock, to the more scholarly but equally gloomy publications of the Club of Rome.[4] But, of course, much else has been written. The hunger of much of the world has been seen as due to a bewildering variety of causes; excessive population growth remains the most popular, but there is no shortage of other explanations. These include the greed of Europeans and their addiction to meat, the evils of colonial and neocolonial exploitation, the incompetence of socialist collective agriculture and the backwardness of traditional farmers; over the last decade some writers believe that environmental degradation, particularly in Africa, has reduced food supplies.

This book deals with what has happened: it does not predict, nor does it claim that there is one overriding cause of hunger. Instead an attempt is made to trace the changes in food consumption which have taken place since the end of the Second World War, or more exactly, because the first statistics became available then, since about 1950.

The world food problem since 1950 is that a large proportion of the population of Africa, Asia and Latin America has been undernourished or malnourished or both. This is a chronic condition. Famines (acute food shortages) that afflict local areas for relatively short periods are not discussed, although it is often outbreaks of famine such as that in the Sahel in 1972–4 and in Ethiopia in 1983–5 and 1991 that draw attention in the West to the problems of the developing countries. Although few deny that there are problems of hunger in the developing world, estimates of the numbers of people undernourished in the late 1970s and the 1980s varied from 62 million to 3000 million (see table 2.1 later). Obviously, the definition and measurement of hunger is far from easy, and this is discussed in chapter 2. The world food problem has not emerged in the post-war period simply as a result of rapid population growth in the developing countries. Indeed, as is shown in chapter 3, the proportion of the world's population with inadequate diets has probably diminished since 1950; prior to the end of the nineteenth century, malnutrition if not undernutrition was as widespread in Europe as in many parts of the developing world today.

Hunger has been attributed to both population growth – the idea that numbers have grown more rapidly than food output – and to poverty. The changes in population and the extent of poverty are outlined in chapter 4, and in chapter 5 the relationship between food output and population growth is examined more closely.

Most of the rest of the book attempts to trace the growth of food output since 1950. The expansion of the world's arable land is considered in chapter 6, and then each of the major regions is discussed – the developed countries, Africa, Latin America and Asia. These chapters try to trace precisely how food output has been increased and to consider some of the difficulties that have been faced and, in some countries, overcome. The penultimate chapter deals with the trade in foodstuffs, for few developing countries are self-sufficient.

This book is not an essay in prophecy; nor does it offer any easy solution to the problems of hunger. It tries to show what has happened in the immediate past. Some lessons can be learnt from this and may be of help in facing the future.

2
The Extent of Hunger

There has been no shortage of attempts to estimate the numbers suffering from hunger. In 1950 Sir John Boyd Orr, the first Director of the Food and Agriculture Organization (FAO), claimed that a 'lifetime of malnutrition and actual hunger is the lot of at least two-thirds of mankind'; the same proportion had been suggested by a League of Nations committee in 1928.[1] In the 1970s and 1980s there were several attempts to estimate the number undernourished or malnourished; these varied greatly (table 2.1) because different authorities defined hunger in different ways. At one extreme T. T. Poleman argued that undernutrition was confined to children under 5 years of age and to pregnant and lactating women; in contrast J. Katzmann classified countries on the basis of the available food supplies per caput. His estimate included the total population of all countries with a daily per caput supply of less than 2900 calories and 40 grams of animal protein. But before examining the ways of estimating the extent of hunger it is necessary to consider some aspects of human needs for food.

UNDERNUTRITION AND MALNUTRITION

Nutrition experts have conventionally distinguished between undernutrition and malnutrition, although the two conditions are interrelated.

The human body needs energy for two purposes. A human being who is at rest and performs no activities at all still needs energy for the brain, heart, lungs and other organs to function. This is the *basal metabolic rate* (BMR). Energy for this and other functions is obtained from eating food that is converted into energy in the body. Most foods have some energy value, measured in calories, but this varies greatly from food to food. Thus 1 ounce of cheese provides 120 calories, but 1 ounce of lettuce only 3 calories.[2] Estimates of the calorie intake needed to

Table 2.1 Estimates of the numbers undernourished in the 1970s and 1980s

Numbers undernourished (millions)	Year	Notes	Source
325–472	1969–71		FAO, 1987
62–309	1975	Number of children under 5 and pregnant women	Poleman, 1981
455[a]	1972–4	Number having available calories less than 1.2 BMR	FAO, 1977
1000	1973	'Suffering from overt hunger'	Borgstrom, 1973
1000	1970	'Serious hunger or malnutrition'	Brown and Eckholm, 1975
1373	1975	–	Reutlinger and Selowsky, 1976
1200–1500[a]	1977	–	Ensminger and Bomani, 1980
1500	1970	–	Berg, 1973
3000	1975–7	–	Katzmann, 1980
335–494[a]	1979–81	–	FAO, 1987
340–730[a]	1980	–	World Bank, 1986
340–512[a]	1983–5	–	FAO, 1989

BMR, basal metabolic rate.
[a] Developing market economies only.
Sources: A. Berg, *The Nutrition Factor: Its Role in National Development*, Washington, D.C., 1973, p. 5; G. Borgstrom, *The Food and People Dilemma*, North Scituate, Mass., 1973, p. 53; L. R. Brown and E. P. Eckholm, *By Bread Alone*, Oxford, 1975, p. 32; D. Ensminger and P. Bomani, *Conquest of World Hunger and Poverty*, Ames, Iowa, 1980, p. 35; FAO, *The Fourth World Food Survey*, Rome, 1977, p. 53; FAO, *The Fifth World Food Survey*, Rome, 1987; FAO, *Current World Food Situation*, Rome, June, 1989; J. Katzmann, 'Besoins alimentaires et potentialities des pays en voie de développement', *Mondes Développées*, 29–30, 1980, pp. 53–6; T. T. Poleman, 'A reappraisal of the extent of world hunger', *Food Policy*, 6, 1981, 236–52; S. Reutlinger and M. Selowsky, *Malnutrition and Poverty: Magnitude and Policy Options*, World Bank Staff Occasional Papers No. 25, Baltimore, Md., 1976; World Bank, *Poverty and Hunger: Issues and Options for Food Security in Developing Countries*, Washington, D.C., 1986.

maintain the BMR have been made, and averages for men and women of different weights and ages have been published. Thus the *average* daily intake necessary for a man weighing 56 kilograms is 1580 calories.[3] Unfortunately for those who wish to estimate the minimum calorific requirements there is evidence of considerable variation between individuals in the metabolic rate, and in addition there is some evidence that the metabolic rate of adults on a low calorie intake adjusts without any change in health or work ability.[4]

But human beings obviously need a calorific intake above the BMR to go about their daily lives. Food is needed for work, but the amount needed varies according to the type of work done. Thus men of a given body weight involved in sedentary activities such as office work expend only 1.8 calories per minute, but a man labouring in the building industry expends 6 calories per minute and a man felling trees expends over 8 calories per minute. An FAO committee has estimated that adult males – of the same weight – require 2700 calories per day where only light activity is undertaken, 3000 calories for moderate activity, 3500 calories for a very active occupation and 4000 calories for exceptional activity.[5] But there are further problems in estimating the average daily calorific needs. The needs of the average woman are less than those of the average man, except during pregnancy and breast feeding; the daily needs of children and adolescents are less than those of adults, except during the spurt of growth before adolescence. With increasing age people work less hard and the metabolic rate declines, and so energy needs are less. Furthermore, food is needed to heat the body, and so calorific needs are greater in cold than in hot climates. To complicate the picture still further there are some studies that show that men with a very low calorific intake can carry out the same work as efficiently as men on much higher calorific intakes without any adverse effects on health.[6]

It is clearly difficult to prescribe the precise calorific intake that is necessary to avoid *undernutrition*. A man who is initially healthy and working satisfactorily can be said to become undernourished if subsequently either his body weight falls or his capacity to work diminishes, or both occur together. A human being deprived of all food will starve to death; this occurs after about forty days without food and when the original body weight has fallen by about 40 per cent.[7]

PROTEINS AND VITAMINS

Human beings who are chronically undernourished are lighter and shorter than those who are adequately fed, and their ability to work will also be less. But the diet may also be deficient not only in quantity but also in quality. This is described as *malnutrition*. The human body needs not only calories for energy purposes but also protein and vitamins; the absence or insufficiency of these nutrients gives rise to specific diseases.

The role of protein in human nutrition has been a matter of much controversy.[8] Protein is needed during growth, and is necessary to replace body tissue. Most plant and animal foods contain protein, but in differing amounts and of different biological value. Protein is made up of a number of amino acids; twelve of these are essential, i.e. they cannot be synthesized in the human body but have to be acquired from foods. Eggs and some other animal foods contain all the essential amino acids; no single plant food, in contrast, contains them all. It was once thought that plant goods contained too little protein to satisfy human needs and that only animal protein would provide the essential amino acids. However, it is now believed that a purely vegetarian diet, if eaten in sufficient quantity and variety, will provide the minimum amount of protein necessary for health; and that although no single plant food will provide all the essential amino acids, these can be provided by eating a mixture of vegetable foods.[9] Most traditional diets do contain the appropriate combination, but there are some exceptions. First, the tropical roots such as manioc and sago have a very low protein content and diets dependent upon these crops may well not provide sufficient protein.[10] Second, as the protein content of most food crops is only about 10 per cent, they have to be eaten in considerable bulk if the minimum protein requirements are to be met; this is no problem for adults, but it may be difficult for infants and young children. Third, animal foods are not only a source of protein but also the major source of some vitamins. Consequently some nutritionists believe that at least 5 per cent of the total protein intake should be of animal origin.[11]

These changed views on the role of protein in the diet have led to changes in policy on nutrition and medical diagnosis. In the 1960s many agencies such as the FAO believed that the fundamental cause of malnutrition was the lack of animal protein, and developing countries were urged to increase their output of animal foods. Now the FAO argues that if sufficient calories are provided the diet will contain enough protein, and indeed enough vitamins. In the 1960s it was thought that protein deficiency diseases such as kwashiorkor were due to an inadequate amount of protein in the food eaten. Subsequent research has shown that in many cases children with kwashiorkor apparently had been receiving an adequate protein intake, but too few calories. Their bodies then used protein stored in muscles and elsewhere as a source of energy. Thus the characteristic symptoms of protein deficiency were due not to a low protein intake in the food supply but to an inadequate energy intake.[12]

The importance of minor elements in the diet has been slowly discovered over the last seventy years, as it has been shown how the absence of vitamins can lead to specific deficiency diseases. An inadequate intake of vitamin A (or retinol) leads to poor sight and eventually may cause blindness; vitamin D deficiencies cause poor bone formation – rickets, common in Britain and the United States in the 1930s, is one consequence. A lack of iodine causes goitre. Vitamin C deficiency is a cause of scurvy, which, however, is now rare. The lack of vitamin B_1 (or thiamine) gives rise to beriberi, a disease found in much of East and South-east Asia – but not India – in the late nineteenth and early twentieth centuries. Most of the vitamin B_1 in rice was contained in the husk, which was removed in milling after the introduction of steel rolling mills in the 1870s. It is a disease now rarely found. Pellagra is more widespread, but on the decline. It is due to a shortage of niacin, one of the vitamin B group, and was found in people living mainly upon maize; once this was realized the incidence of the disease declined and it is now only common in Africa.[13]

Although vitamin deficiency diseases – together with others such as that caused by lack of iron – are still widespread, they have nearly everywhere declined over the last half century.[14] The principal reason for this is that the amounts of vitamin needed to eliminate the diseases are very small and they can all be synthesized in the laboratory. In contrast the provision of energy and protein still requires considerable quantities of foodstuffs.

MALNUTRITION AND INFECTIOUS DISEASE

Although it might seem self-evident that a malnourished person should be more susceptible to infectious disease, in the immediate post-war period this was not thought to be so. Studies of the Dutch population in the famine of 1944 showed that there was no increase in the prevalence of infectious disease, and similar conclusions were drawn from other work on starvation and disease.[15] However, views on the relationship between infectious disease and malnutrition have changed.[16] It is now believed by some that malnutrition damages the body's immune systems, and that malnourished people are more likely to catch certain infectious diseases and are less likely to survive them than a well-fed person. Furthermore, the prevalence of infectious disease is thought to increase malnutrition. Malnutrition is found in areas of poverty, which are also areas of poor hygiene and sanitation. In such areas diseases of the stomach and intestines are common, to which infants and young children are particularly susceptible. Such diseases cause a loss of appetite and vomiting and impair the ability of the gut to absorb nutrients. Thus children who are offered sufficient food, or indeed eat it, none the less may suffer from malnutrition because of their poor health. Thus it is now agreed that the improvement of public and private sanitation and hygiene would reduce the prevalence of malnutrition among children, and that malnutrition is by no means always simply a consequence of inadequate food supplies. The post-war history of China illustrates this point. Although food supplies per capita changed little between the 1950s and the late 1970s, the average weight and height of children and adolescents increased. This has been attributed to the decline of intestinal diseases.[17]

ESTIMATING THE NUMBERS MALNOURISHED AND UNDERNOURISHED

There are three ways in which the numbers suffering from hunger can be estimated: through the symptoms of deficiency diseases; through the amount of food consumed; and through the amount of food available.

The symptoms of malnutrition and undernutrition

Undernutrition and malnutrition give rise to diseases which can be recognized by the clinical diagnosis of symptoms; by the measurement of body weight, height and other indicators and their comparison with healthy growth; and by changes in the chemical composition of the body.[18] The only accurate way to measure the extent of hunger is by counting the number of people with such symptoms. This is unfortunately difficult for several reasons.

First, much of the population of the developing world lacks medical care and nutritional experts or teams capable of undertaking such inquiries. Although there are numerous published accounts of the diagnosis and treatment of nutritional diseases, or measurements of changes in weight and height, they cover a very small proportion of the world's population. Furthermore many such surveys have been undertaken during periods of food crisis, and their results may not be representative of the population under normal conditions.

Second, most of the diseases due to poor nutrition may occur in both extreme and mild forms. Protein calorie malnutrition, also known as protein energy malnutrition, which afflicts principally young children, may occur as kwashiorkor, which can lead to death if not treated; it may also be used to describe children slightly less in height and body weight than well-fed children. The magnitude of the problem depends upon the diagnostic criteria adopted. Thus a survey of the health of children in a number of countries in Central America found that between 0.1 and 0.9 per cent were suffering from severe forms of protein energy malnutrition, but between 50 and 73 per cent had reduced weight for their age. Similarly, in India it was estimated that 1.2 per cent of children had kwashiorkor, but 80 per cent showed signs of reduced growth due to an inadequate diet.[19]

Thus information upon the prevalence of deficiency diseases is sporadic and it is not possible to estimate the numbers suffering from hunger in this way. Such information that is available is summarized here.

Of the vitamin deficiency diseases, scurvy, which is due to a shortage of vitamin C and was once common in parts of Europe, is now rare in both the developed and the developing world. Rickets, still widespread in Western Europe and the United States in the 1930s, has greatly declined, although it is still found in parts of the tropics. Pellagra has also declined, but is still found. It is due to a lack of niacin, one of the vitamin B group, and was found in those who obtained a high proportion of their diet from maize. However, pellagra has never been common in Central and South America, where maize was domesticated and is widely grown. Here it was invariably eaten with beans, and the corn was treated with alkali before grinding. Pellagra was first reported in southern and eastern Europe after the introduction of maize and its widespread adoption as a food crop in the eighteenth century. It is still found in Romania, but is unknown elsewhere in Europe. The major incidence of the disease is now in Africa, where it has also been found among those dependent, not upon maize, but sorghum.[20] Beriberi has greatly diminished since its cause was discovered (see p. 9). Two diseases have not declined. In 1958 it was estimated that there were 200 million cases of endemic goitre in the world, caused by lack of iodine. Xeropthalmia, due to an inadequate amount of vitamin A, damages the eye; it has been estimated that there are 700,000 new cases a year. Some 3 million children under 10 in the developing countries are blind.[21]

Of greater significance than these diseases are the illnesses and deaths that arise among children who are lacking in calorie or protein supply. Children who are breast fed do not suffer from malnutrition, but once they are weaned they are very much at risk. First of all, infants in impoverished areas are highly susceptible to disease. In a study of children in a Mexican village it was found that on average each child had thirty-five illnesses in the first two years of life, and was ill for one-third of the time; this would reduce the efficiency of food intake considerably.[22] Second, although the absolute amounts of calories and protein needed by young children are less than those of adults, the needs per unit of

Table 2.2   Range and median of percentage prevalence of protein calorie malnutrition in community surveys, 1963–72

Area	Number of surveys[a]	Number of children examined	Severe forms Range (%)	Severe forms Median (%)	Moderate forms Range (%)	Moderate forms Median (%)
Latin America	11	108,715	0.5–6.3	1.6	3.5–32.0	18.9
Africa	7	24,759	1.7–9.8	4.4	5.4–44.9	26.5
Asia[b]	7	39,494	1.1–20.0	3.2	16.0–46.4	31.2
Total	25	172,948	0.5–20.0	2.6	3.5–46.4	18.9

[a] Surveys were all of at least 1000 children, mainly under 5 years old.
[b] Excluding China and Japan.
Source: J. M. Bengoa, 'The state of world nutrition', in M. Rechcigl, Jr. (ed.), *Man, Food and Nutrition*, Cleveland, Ohio, 1973, p. 6

body weight are high. Studies of food consumption within households are rare, but suggest that women and children get not only less than adult males but also less than their needs. Poorly fed children are often permanently affected by inadequate food intake. Their height and body weight are retarded, and as adults they are shorter and lighter than the well fed; some have suggested that malnutrition in childhood permanently retards mental development.[23]

The acute forms of disease resulting from hunger are nutritional marasmus and kwashiorkor. In the former, muscles waste, there is a loss of subcutaneous fat, the buttocks diminish, the skin is loose, and there are eye lesions and skin rashes. The child is weak, apathetic and tires very easily.[24] This is due to a lack of calories. In contrast, protein deficiency causes kwashiorkor, where hair reddens and straightens, the body is typically swollen, the face is moonlike in shape, and skin rashes and ulcers occur.[25] Many children exhibit symptoms of both diseases, and it is now usual to refer to *protein calorie malnutrition*. As noted earlier, this can vary from the acute stage of kwashiorkor or marasmus to mild retardation of growth. J. M. Bengoa reviewed the results of surveys of protein calorie malnutrition in forty-six countries between 1963 and 1972 (table 2.2) and has also estimated the absolute numbers of children under 5 years of age suffering from malnutrition (table 2.3), apparently by assuming that the *median* figure for each continent in table 2.2 can be used as a *mean* to calculate the percentage of the numbers under an

Table 2.3 Estimates of the total numbers of children under 5 years of age suffering from severe or moderate protein calorie malnutrition, *c*.1970

Area	Severe (thousands)	Moderate (thousands)	Total (thousands)
Latin America	700	9,000	9,700
Africa	2,700	16,000	18,700
Asia[a]	6,600	64,000	70,600
Total	10,000	89,000	99,000

[a] Excluding China and Japan.
Source: J. M. Bengoa, 'The state of world nutrition', in M. Rechcigl, Jr. (ed.), *Man, Food and Nutrition*, Cleveland, Ohio, 1973, p. 7

age of 5 in 1970. As the population of less developed countries – excluding China and Japan – increased by 48 per cent between 1970 and the late 1980s, a similar incidence of malnutrition would yield about 190 million children under 5 suffering from malnutrition.

Bengoa has also made an estimate of the number of children under 14 years of age suffering from protein calorie malnutrition. In 1966, 269 million from a total of 667 million – 40 per cent of children under 14 – were suffering from undernourishment. There are no comparable figures for more recent years, but in 1989 approximately 1429 million people in the developing world were under 15. If the incidence of protein malnutrition was the same as in 1966 then 570 million would have been affected by this disease.[26] Two recent surveys of sample studies of the weight and height of children in developing countries have produced very similar results. A survey of studies made around 1980 concluded that 39 per cent of under 5s were 80 per cent below the standard weight for height in developing countries, a total of 141 million.[27] Some later studies (table 2.4) indicate that about 40 per cent of the under 5s in developing countries (excluding China) had stunted growth and were underweight for their age, a total of 149 million.

The consequences of malnutrition among preschool children include stunting of growth, low body weight and listlessness, which is reflected in adulthood in a reduced capacity to work. But malnutrition may also be directly or indirectly responsible for the deaths of children under 5. A high proportion of children of this age in areas of malnutrition are also victims of numerous respiratory and intestinal infections and this makes the cause of death difficult to determine. However, an analysis of some 11,000 deaths of children under 5 in Latin America in the

Table 2.4 Children suffering from symptoms of undernutrition, 1980s

	Sub-Saharan Africa	Middle East and North Africa	Asia and Oceania	South Asia	East and South Asia	Latin America and the Caribbean	All developing countries	Industrial countries
Percentage with low birth weights	16	7	19	28	9	11	17	6
Percentage of under 5s underweight	29	–	45	45	43	14	38	–
Percentage of 1–2 year olds wasted	16	–	16	18	–	3	13	–
Percentage of 2–5 year olds stunted	38	–	50	50	–	31	42	–

Source: United Nations, *Human Development Report 1990*, Oxford, 1990

1960s found that malnutrition was the underlying cause of death of 7 per cent and an associated cause for 46.2 per cent.[28]

Household surveys and income distribution studies

The only accurate way to measure hunger is to count the numbers with symptoms of nutritional deficiency diseases. As has been seen this is difficult and the estimates available refer mainly to children. A second but indirect approach is to measure the amount of food consumed by individuals, by households or by groups, and to compare this with some estimate of the minimum requirements needed to avoid malnutrition and undernutrition. Both these approaches have limitations.

Studies of food intake necessarily deal only with comparatively small groups. The most accurate way to estimate the nutrients received is to measure the food before it is digested and to convert this value. Most such surveys are based upon answers to questions about the food consumed by households in the preceding day or in the preceding week. Such surveys are difficult to organize, rarely cover the whole of a year and may not be representative of the total population of a country. An alternative approach is not to measure calorific intake but to record the amount of money spent on food, to determine the cost of a minimum diet and then to estimate the numbers who cannot afford this.[29]

Both these methods require the measured intake to be compared with a norm, a safety level or the minimum requirements. Over these there has been much controversy. Various FAO and World Health Organization (WHO) committees have published recommended levels in terms of calories and protein. They have begun by recommending the minimum needs for an adult male aged 20–39 with a given body weight and a given level of activity. Tables are also prepared to show the reductions or increases necessary for age, sex, body weight and different levels of activity. The FAO recommendations published in 1973 were used in estimating the incidence of hunger in the *Fourth World Food Survey* and recent World Bank estimates, but have been superseded by a more recent report. The 1973 report was criticized principally on the grounds that the average requirements will conceal a great range in individual needs.[30] P. V. Sukhatme has pointed out that the FAO average for an adult male in India is 2550 calories, but individual minimum needs may range from 1750 to 3350 calories. It seems generally agreed that the FAO requirement figures overstate minimum needs, and thus any estimates based on these data will overestimate the extent of hunger. Certainly some household surveys demonstrate that some groups live healthy, hard working lives on calorie intakes much lower than the FAO requirements. Thus tribes in New Guinea led healthy lives on 1700 calories per day, and in Central Java apparently healthy villagers only received 1392 calories per day. On the other hand labourers in Panama, to whom food was made freely available, ate an average of 3355 calories per day without any sign of gaining weight.[31]

Quite apart from the methodological problems of interpreting such surveys, they are not sufficiently representative of national populations to allow any national or world estimates of hunger to be made. Much the same may be said of income surveys: however, surveys of income and food consumption in various countries do reveal facts of major importance about the consumption of food. Sample surveys of households in Bangladesh, the Philippines, Malaysia and Guatemala in the 1980s (figure 2.1) show that there was a considerable range in average expenditure upon food per caput, and that this was closely related to income. In Bangladesh expenditure upon food in the richer families in the sample was five to six times that in the poorest. Not surprisingly richer groups bought more meat than poorer groups; more surprisingly in Guatemala the expenditure on cereals as well as meat increased with incomes. Such data support the assumption that poverty is the main cause of hunger, although it should be remembered that in many developing countries much of the food for farm families is produced on the farm and income levels may be irrelevant. Surveys of calorie consumption by households are less common, but indicate a way of measuring the number of people who have less than the minimum requirements. Thus in India in the early 1970s 58 per cent of the sample population had less than the national mean of 2217 calories, and 79 per cent had less than the FAO adult male's minimum requirement for India (table 2.5).

A further problem in interpreting household consumption surveys is that calorie intake is normally measured as per caput for the household. Yet there is evidence that there is maldistribution within most households, with men getting more than women. Children – except in early adolescence – need fewer calories than adults, but a study of peasant households in Nigeria showed that although adults were getting more than the FAO requirements, children were getting considerably less than the estimated needs of their age group. Surveys of household consumption in Bangladesh showed that boys got 16 per cent more food than girls, men 29 per cent more than their wives.[32]

There is of course a considerable difference in expenditure – in real terms – upon food between the developed and the developing countries. The developing countries' expenditure is less than half in most countries for which data are available, except in Latin America, and less than one-fifth in sub-Saharan Africa and India, the two regions where undernutrition is most prevalent (table 2.6).

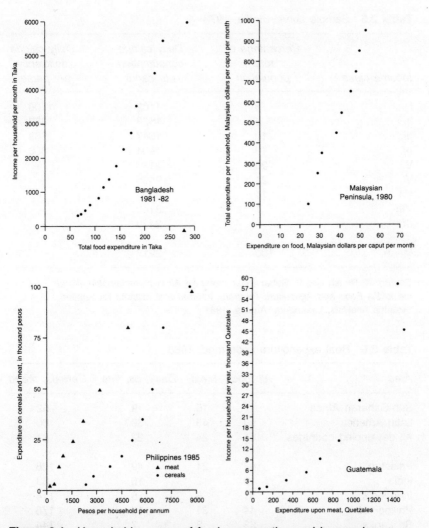

Figure 2.1   Household surveys of food consumption and income in
Bangladesh, Malaysia, the Philippines and Guatemala in the 1980s
Source: FAO, *Review of Food Consumption Surveys – 1988*, Food and Nutrition
Paper 44, Rome, 1988

Food availability data

A third way in which the extent of hunger can be measured is to
estimate the amount of food available in a country in a given year, and
to express this in terms of calories per caput per day. The FAO has
published the results of such food balance sheets annually for all countries

Table 2.5 Sample survey, India, 1973–4

Income class	Percentage of total population	Daily calorie consumption per caput	Daily calorie deficiency per person
I	5	1102	1108
II	5	1528	682
III	10	1647	563
IV	18	1904	306
V	20	2115	–
VI	21	2495	–
VII	11	2805	–
VIII	7	3140	–
IX	3	3440	–
Mean	100	2217	–

Source: K. Parikh and F. Rabar (eds), *Food for All in a Sustainable World: the IIASA Food and Agriculture Program*, International Institute for Applied Systems Analysis, Laxenburg, Austria, 1981

Table 2.6 Real expenditure upon food, 1988

Area	All food	Meat	Dairy, oil, fats	Cereals, bread
Sub-Saharan Africa	17	15	9	52
Latin America	59	48	36	80
All developing countries	41	24	22	87
Pakistan	42	21	49	128
India	19	2	16	63
Indonesia	31	6	5	114
Philippines	44	31	10	175
Sri Lanka	33	3	7	144

Developed countries = 100.
Source: United Nations, *Human Development Report 1990*, Oxford, 1990, p. 172

since 1960, and for a selection of countries in the 1930s and for the period 1948–52. Most estimates of the extent of world hunger are based upon these figures.[33]

A figure for an individual country is arrived at in the following way. The total food output in a year is estimated and converted into calories. There is deducted from this all exports of food crops, crops fed to livestock, the food used for industrial purposes and seed for the following year's crop. Added on to this figure is the calorific value of food imports and any food in stocks from previous years. It is assumed that there is a loss due to pests and diseases between field and retail outlet, and so the total output is reduced by 10 per cent.

These estimates have been much criticized.[34] Few developing countries have annual agricultural censuses, and so for many of these annual food output is estimated by the FAO statisticians. Even where censuses are carried out farmers are notoriously reluctant to report their output accurately for fear of taxation, and yields tend to be underestimated. Output of meat is difficult to measure unless there are figures from slaughterhouses. The estimates exclude alcoholic drinks, game, human milk and the gathering of fruits and berries, and do not allow for any loss in preparation in the kitchen or in waste. Thus even as an estimate of total calories available per caput per day these figures have their defects. In 1986–8 the figures ranged from 3901 calories in Belgium to only 1604 calories in Mozambique. The average daily supply in the developing countries was 3398 calories, in the developing only 2434 calories.

That there is a difference in food availability between the developed and the developing countries is not in question. However, the contrast is not quite as simple as averages for the two blocs would suggest. Although of the developed countries only Japan has a figure of less than 3000 calories per caput per day (figure 2.2), there are several countries in the developing world with food supplies little short of the developed countries, particularly in the Middle East and South America. There is indeed a continuum between those with high and those with low supplies, as can be seen when countries are arranged in order of the available calories per caput per day (figure 2.3).

The available food supply data has been used by many writers on the assumption that it is identical to consumption. This is not so. In some developed countries annual records of food consumption are kept. The United Kingdom annual food survey illustrates some of the problems of interpreting food availability data. Each year the United Kingdom government collects data on quantities of food purchased for consumption in a sample of households. It can be seen (table 2.7) that average per caput consumption by households reached a peak in 1954–6 and has since declined; in 1986–8 it averaged only 2366 calories but the FAO available food supply estimate was 3259 calories. This discrepancy is partly explained by the fact that household consumption excludes sweets, soft drinks and alcohol, which provided 330 calories per day, and does not include meals eaten outside the household, which have increased. On the other hand, no estimate is allowed in the household

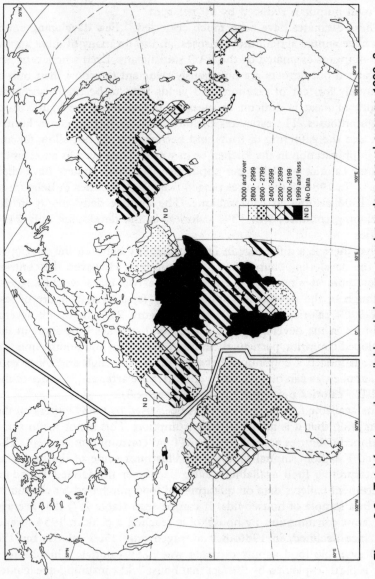

Figure 2.2 Total food supply, available calories per caput per day, annual average 1986–8

Sources: FAO, *Production Yearbook 1989,* vol. 43, Rome, 1990; *Food Balance Sheets 1986–88,* Rome, 1990

3000 and over
2800 - 2999
2600 - 2799
2400 - 2599
2200 - 2399
2000 - 2199
1999 and less
N D No Data

Total calories per caput per day

Figure 2.3 Calories per caput per day, by countries, 1986–8
Sources: FAO, *Production Yearbook 1989*, vol. 43, Rome, 1990; *Food Balance Sheets 1986–88*, Rome, 1990

Table 2.7 Estimates of consumption and available food supplies, UK, 1948–88

Period	Foodstuffs moving into consumption	FAO available food supplies	Household consumption survey
1948–50	–	3130	2474[b]
1951–3	–	3110	2479
1954–6	3130[a]	3220	2660
1957–9	3153	3230	2623
1960–2	3170	3270	2633
1963–5	3156	3260	2613
1966–8	3096	3180	2570
1970–1	3072	3170	2525
1972–4	3020	3376	2383
1975–7	2923	3305	2276
1978–80	2906	3255	2246
1986–8	–	3259	2366

[a] 1956 only.
[b] 1950 only.
Source: HMSO, *Domestic Food Consumption and Expenditure, Annual Reports*, London, 1950–88; FAO, *Production Yearbooks*, Rome, 1966–89

consumption figure for waste in the home. But the annual report does also include an estimate of the amount of food moving into consumption, which includes foodstuffs going to retailers, institutions, restaurants, manufacturers of soft and alcoholic drinks and of sweets (table 2.7). These estimates are nearer the FAO figures, although consistently lower. Similar discrepancies have been found in other countries. A survey of food consumption in the United States in 1976 and 1980 found that the average daily intake for males was 2381 calories, for women 1579 calories. The available food supply calculated by the FAO for these two years was 3552 and 3652 calories per day respectively. Such differences between consumption and availability data have also been found in Japan and in a number of developing countries.[35]

ESTIMATES OF THE NUMBERS UNDERNOURISHED

Although the very low figures for available supply per caput in much of Africa, South Asia and some other parts of the developing world suggest that food consumption there must be low (figure 2.2), as they stand they give no indication of the distribution of hunger. However, these figures have been used by the FAO and the World Bank to estimate the extent of hunger more precisely. There are two ways in which this can be done. First, the national availability and national requirements can be compared; and second, the food availability data can be combined with figures on income distribution to estimate the numbers under-nourished.

National needs

FAO statisticians have calculated the number of calories needed per caput per day to provide everyone in the country a diet which would prevent undernutrition, assuming that the calorie supply available was distributed according to the needs of each individual, which depend upon age, sex and activity level.[36] A national requirement was produced for every country, ranging from 2000 calories to 2710 calories per caput per day. National requirements for developing countries are lower than for developed countries because a greater proportion of the population are children.[37]

Food availability can then be expressed as a percentage of minimum requirements (figure 2.4). This shows that if food supplies were distributed according to individual needs there would be no undernutrition anywhere in North America, Europe, the USSR, Australasia or Japan. This is much as expected. But in 1986–8 comparatively few developing countries had food supplies – including production, stocks and imports – *less* than national requirements. The areas where food supplies remain inadequate are tropical Africa and South Asia; in contrast thirty years ago the great majority of developing countries had supplies less than requirements (see figure 3.6 later).

Income distribution and consumption

There are undoubtedly problems of malnutrition in many of the developing countries that have supplies which are above requirements. The reason for this, of course, is that food supplies are not allocated according to individual need, and it is unlikely that there is any country where this is so. Even where rationing is practised – as in China in the early 1970s – malnutrition occurs. Chinese officials subsequently admitted that at least 10 per cent of the population then had inadequate diets. In the mid 1980s, after agricultural reforms had greatly increased output, 11 per cent of the rural population were thought to be undernourished.[38]

The main reason for the great differences in calorific intakes between

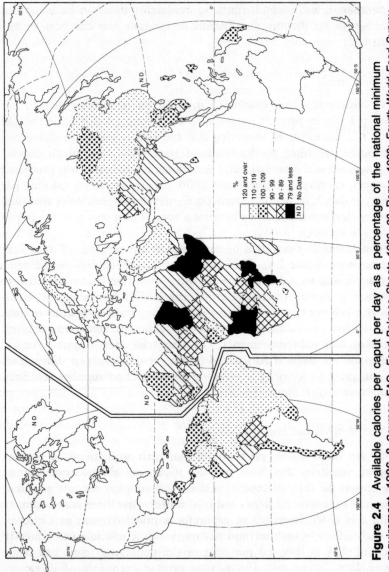

Figure 2.4 Available calories per caput per day as a percentage of the national minimum requirement, 1986–8 *Sources: FAO, Food Balance Sheets 1986–88, Rome, 1990; Fourth World Food Survey, Rome, 1977*

different groups in the developing countries is income, as it is indeed in the developed countries. The difference is that in the developed countries the lowest income groups have, generally, sufficient money to buy an adequate diet. In the United Kingdom there are only minor differences in calorie and animal protein consumption between income groups. In the developing countries this is not so, as can be seen from figure 2.1 and table 2.5. In India in the early 1970s 5 per cent of the sample population had an average daily consumption of only 1102 calories and 20 per cent had less than 1700 calories, neither of which could have produced an adequate diet.

World Bank and FAO estimates

Because of the importance of income distribution, officials of FAO and the World Bank have produced independent estimates of the extent of undernutrition which combine the FAO food availability data and income distribution figures. Their estimates for the 1970s and the 1980s are reproduced here.

The *Fourth World Food Survey*, published by the FAO in 1977, provided estimates for the numbers undernourished in 1972–4. This calculation was based upon estimates of the average daily consumption and, by making assumptions about the distribution of income, the proportion of the population of a country receiving a given number of calories per day. The minimum number of calories required to avoid undernutrition was set at 1.2 times the BMR and the numbers receiving less than this amount were estimated for each country. As J. C. Waterlow observed, if people are existing at this level – and are clearly not totally inactive all day – then they must have made a remarkable metabolic adaptation to such low calorific intakes.[39] Furthermore the estimate excluded China. If an estimate for China – 10 per cent of the rural population – is included, then in 1972–4 530 millions in the developing countries suffered from undernutrition (table 2.8).

In the 1970s the World Bank published an estimate of the numbers malnourished in 1975. This used the 1965 data on food availability per caput per day for developing countries – but again excluding China – and assumed that the FAO national requirements per caput were the minimum level needed to avoid undernutrition. These figures were of course much higher than the FAO minimum requirement of 1.2 BMR. Assumptions were made about income distributions within each country, and the average daily consumption of each income group was calculated. The numbers falling below minimum requirements were then calculated, and projected forward to 1975 using assumed rates of population income and food output growth. This work concluded that

Table 2.8 Number of persons with food intake below the critical limit, 1972–4

Area	Percentage below 1.2 BMR	Number below 1.2 BMR (millions)
Africa	28	83
Far East[a]	29	297
Latin America	15	46
Near East	16	20
Developing countries	25	455

BMR, basal metabolic rate.
[a] Excluding China.
Source: FAO, The Fourth World Food Survey, Rome, 1977, p. 53

Table 2.9 Numbers and proportion undernourished, 1975

	Millions	Percentage of population
Latin America	112	36
Asia[a]	924	82
Middle East	94	51
Africa	243	77
Developing countries	1373	71

[a] Excluding China.
Source: S. Reutlinger and M. Selowsky, Malnutrition and Poverty: Magnitude and Policy Options, World Bank Staff Occasional Papers No. 23, 1976, p. 31

71 per cent of the developing countries population was undernourished in 1975 (excluding China) (table 2.9). After some criticisms this was later revised downwards to between 40 and 60 per cent of the population of the developing world. Even so this revised total was well above the FAO estimate for 1972–4.[40]

In 1987 the Fifth World Food Survey was published by the FAO, presenting estimates of the numbers undernourished in 1979–81 and also in 1969–71. More recently estimates on a similar basis have been published for 1983–5. The methodology used in this survey differed substantially from that in the Fourth World Food Survey. An attempt has been made to estimate the number of households receiving less than 1.2 or 1.4 BMR, using the food availability data and income distribution information. Again China is excluded from the estimates. In 1983–5 (table 2.10) 348 million, 14.6 per cent of the population of the developing

Table 2.10 Estimates of undernutrition in developing market economies

	Percentage of population			Number of persons (millions)		
	1969–71	1979–81	1983–5	1969–71	1979–81	1983–5
Below 1.2 BMR						
Africa	23.5	21.9	26.0	63	78	105
Near East and North Africa	15.7	6.7	5.6	28	16	15
Asia	19.5	15.6	14.3	190	191	191
Latin America	12.7	9.8	9.5	35	35	37
All developing countries	18.6	14.7	14.6	316	320	348
Below 1.4 BMR						
Africa	32.6	30.6	35.2	86	110	142
Near East and North Africa	22.5	10.8	9.1	41	25	24
Asia	28.7	23.5	21.8	281	288	291
Latin America	18.5	14.6	14.2	51	52	55
All developing countries	27.0	21.8	21.5	460	475	512

BMR, basal metabolic rate.
Source: N. Alexandratos (ed.), *World Agriculture: towards 2000, an FAO Study*, London, 1988, p. 66.

world, received less than 1.2 BMR, 512 million, or 21.5 per cent of the population, received less than 1.4 BMR.

In 1986 the World Bank published a second estimate of the extent of hunger, this time for 1980, again based upon income. It attempted to measure the number of households that did not meet the minimum calorie intake defined by the FAO and the WHO in 1973, and recognized two degrees of undernutrition: those who did not have sufficient calories for an active working life, defined as receiving below 90 per cent of the FAO/WHO minimum requirement; and those who did not receive sufficient calories to prevent stunted growth and serious health risks, defined as being below 80 per cent of the FAO/WHO minimum requirements (table 2.11). These give estimates for 1980 for the developing countries (excluding China) of between 340 and 730 million, compared with FAO estimates of between 320 and 475 million for 1979–81.

Table 2.11 Prevalence of energy-deficient diets in developing countries, 1980

	Below 90% of requirement		Below 80% of requirement	
	Percentage of population	Population (millions)	Percentage of population	Population (millions)
Sub-Saharan Africa	44	150	25	90
East Asia[a]	14	40	7	20
South Asia	50	470	21	200
Middle East and North Africa	10	20	4	10
Latin America	13	50	6	20
All developing countries	34	730	16	340

[a] Excluding China.
Source: World Bank, *Poverty and Hunger: Issues and Options for Food Security in Developing Countries*, Washington, D.C., 1986, p. 17

CONCLUSIONS

It will be apparent that there are great differences in the estimates of the numbers undernourished made by the FAO and the World Bank in the 1970s and 1980s, and this has led some to believe that it is impossible to calculate with any accuracy the numbers undernourished. This is almost certainly true.[41] But the figures do have value for they give some indication of where the greatest proportion of the population are undernourished, and where the absolute numbers are greatest.

In the early 1980s between 320 and 730 millions were undernourished, between 14 and 34 per cent of the population of the developing world. This excludes China, which in the mid 1980s may have had some 87 million suffering from undernutrition.[42] There are important consistencies in the pattern of distribution. In all but one (South Asia in the first column of table 2.11) of the estimates shown, Africa has the highest *proportion* of the total population undernourished, the Near East and Latin America the lowest. However, a quite different pattern emerges if the total numbers undernourished is considered, for it is Asia, and in particular South Asia, where a large proportion of the total numbers malnourished are to be found. Indeed, whichever year or method is considered, at least 60 per cent of the undernourished are in Asia.

There is a further reason to attempt to calculate the numbers under-nourished, and that is to determine whether the problem is getting worse or better. Recent FAO estimates give some indication of this (table 2.10), but before considering these figures the problem of hunger should be seen in a longer historical perspective.

3

A Short History of Hunger

Few of the recent estimates of the extent of world hunger have attempted to measure change over time; even fewer have considered the extent of hunger in the developed world. Yet half a century ago most discussions of world hunger were concerned as much with the problems of malnutrition in Europe and North America as in Africa, Asia or Latin America. Indeed it is only since the end of the Second World War that malnutrition has ceased to be a major problem in Europe and the United States. In this chapter the decline of undernutrition and malnutrition in Europe is described; and then an attempt is made to trace the changes in the extent of hunger in the developing world since the 1930s.

THE ELIMINATION OF HUNGER IN EUROPE

Pre-industrial Europe

Most historians would agree with Carlo Cipolla that before 1800 the poor of Europe – most of the population – were in a chronic state of undernourishment.[1] Those who relied upon wages had to spend most of their income on food. Moreover, because bread was the cheapest source of energy, it and other vegetable foods were the bulk of the diet, with little meat, milk or cheese being eaten. Nor were farmers – at least those with little land – much better off. Wilhelm Abel has calculated that a twelfth-century farmer with six hectares would have to give half his grain to the Church or lord, leaving a family of five with only half a kilogram of grain per head per day – too little by modern standards. Braudel has surmised that pre-industrial Europeans got about 2000 calories per day, much the same as Chinese agriculture produced before the eighteenth century, and below the level of all but a few developing countries today.[2]

But more attention has been paid to the fluctuations in food consumption in pre-industrial Europe than to average conditions, for it was the extremes that attracted contemporary writers and, later, historians. Long- and short-term fluctuations can be distinguished. The long term has been outlined by Braudel and Abel. The bubonic plague of 1348–50 drastically reduced the population of Europe, which did not recover until the late fifteenth century. However, for the surviving farmer there was more land, and, for the labourer and artisan, higher real wages. The population of Europe was better fed in the fourteenth and fifteenth centuries than in the sixteenth, seventeenth or eighteenth centuries. There is little evidence on food consumption to support this; it is inferred from the trends in the purchasing power of wages, which fell in the late sixteenth century and did not recover to their medieval level until the mid nineteenth century. Some fragmentary data on meat consumption suggest that meat was a common part of the diet in the fourteenth century and then declined, not recovering to the medieval level in Germany until the mid nineteenth century.[3]

Rather more attention has been paid by historians to short-term fluctuations in food consumption. They were of two types. First were famines, where food shortages occurred for one or more years over a wide area. Second were the more localized food shortages which occurred only in one parish or a group of parishes, which led perhaps to a few deaths, and which were treated by contemporaries as part of the normal course of events. Indeed these subsistence crises received little attention from historians until work by historical demographers on parish registers revealed their existence.

Famines are a much discussed part of European history. Thus few parts of Western Europe escaped the great crisis of 1314–16, and they are well attested from then until the 1840s. Although it was Ireland that suffered most from the failure of the potato crop in the 1840s, other parts of Europe also had acute food shortages at that time. But this may be described as the last major famine in Western Europe. The frequency and extent of famine had been in decline in Western Europe long before the mid nineteenth century. The last major famines occurred in England in the 1620s, in Scotland in the 1690s, in Germany, Switzerland and Scandinavia in 1732, and in France in 1795, although many parts of Europe were afflicted with harvest failure and high prices in 1816.[4] These major outbreaks, covering large areas, declined in impact in the eighteenth century. Not only were better farming methods increasing crop yields, but new crops like the potato were providing more secure supplies, and a wider range of crops was freeing the population from the hazards of depending upon one grain. Equally important, transport improvement both internally and internationally was reducing isolation, and governments both local and national were paying more attention to what would nowadays be called famine relief.[5]

These major crises have of course received much attention from historians. But there were other less dramatic but more frequent and localized demographic crises which have been revealed by the work of historical demographers studying parish registers, particularly in France. From the late sixteenth to the early eighteenth century, crises were common. Harvest failure led to a rise in the death rate, a fall in the birth rate and fewer marriages. Once the crisis was past, marriages and births recovered. Initially these events were described as *subsistence crises*, and deaths were attributed to starvation. But others have argued that this was rare, and the malnourished were simply more susceptible to infectious disease. As in the developing countries today, poverty, malnutrition and infectious disease were interlocked and interrelated. But even these local crises were diminishing in the eighteenth century: thus although the English rioted over food prices in the later eighteenth century, few died from hunger.[6]

There are no estimates of national consumption levels or available food supplies before the nineteenth century, and the level of food consumption has been inferred from trends in real wages or contemporary descriptions of meals consumed by the very rich or the very poor. In the last two decades historians have attempted to establish the history of European nutrition upon a more scientific basis using the records of food bought for institutions such as monasteries, workhouses, schools and military establishments.[7] However, the national food supplies of various countries have been calculated from the early nineteenth century and converted into their nutritional equivalents. It is thus possible to trace with more confidence changes in the level and composition of national diets from the early nineteenth century to the present.[8]

Trends in national food supplies after 1800

Estimates have been made for Norway from the eighteenth century to the second half of the nineteenth century (table 3.1), for France and Germany from 1800 to the present and for Italy, the United States and Japan from the later nineteenth century (figure 3.1). Together with more detailed estimates for France, they allow the major trends in available food supplies – but not actual consumption – to be discerned.

In the early nineteenth century, Norway (table 3.1) and France and Germany (figure 3.1) had daily food supplies below 2000 calories per caput, a level found in the mid 1980s only in Bangladesh, Haiti and the poorer African countries, and well below nearly all of modern Latin America, North Africa, the Near East and East Asia. France's supply

Table 3.1 Estimated food supplies in Norway (calories per caput per day)

Year	Food supply
1723	1400
1809	1800
1835	2250
1855–65	3300

Source: M. Drake, 'Norway', in W. R. Lee (ed.), *European Demography and Economic Growth*, 1979, p. 293

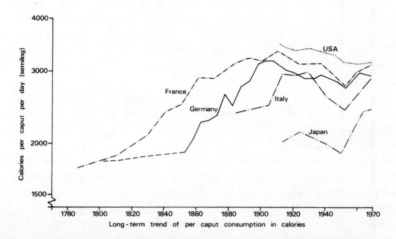

Figure 3.1 Food supply in selected developed countries, 1780–1970
Source: A. Weber and E. Weber, 'The structure of world protein consumption and future nitrogen requirements', *European Review of Agricultural Economics*, 2, 1974–5, pp. 169–92

of total protein in the early nineteenth century – just over 50 grams per caput per day – was also below that of most countries in Africa and Asia today, as was the supply of animal protein. In France and Germany the latter was below 20 grams, again similar to many developing countries at present (figure 3.2). Not all Western Europe was so poorly provided in the early nineteenth century; in the southern Netherlands the daily supply was 2200 calories, but with alcohol included this exceeded 2500 calories; 73 grams of protein were available.[9]

However, after 1800 there was a continuous increase in the national calorie supply per caput in France and Germany (figure 3.1). It exceeded 3000 calories per caput in Norway by the 1850s, in Britain

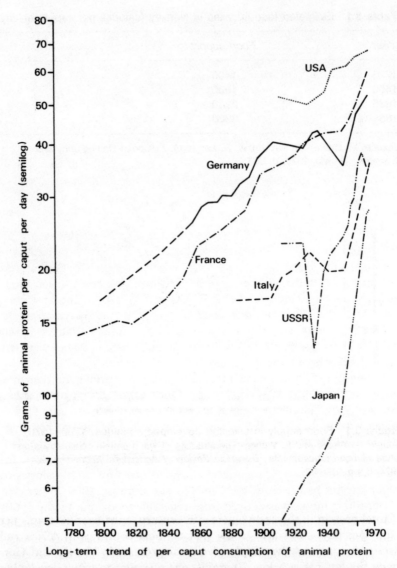

Long-term trend of per caput consumption of animal protein

Figure 3.2 Supply of animal protein in selected developed countries, 1780–1970 *Source*: A. Weber and E. Weber, 'The structure of world protein consumption and future nitrogen requirements', *European Review of Agricultural Economics*, 2, 1974–5, pp. 169–92

Table 3.2 Meat consumption in the German Empire in the nineteenth century (kilograms per head per year)

Year	Meat consumption
1816	13.6
1840	21.6
1873	29.5
1892	32.5
1904–5	46.8

Source: W. Abel, *Agricultural Fluctuations in Europe. From the Thirteenth to the Twentieth Centuries*, London, 1980, p. 267

possibly by the 1850s and certainly by the 1880s, in France by the 1870s and in Italy by the early twentieth century. At the beginning of the nineteenth century not only were national supplies per caput very low, but most of the calories were derived from vegetable foods; three-quarters of the French food supply was provided by bread and potatoes. After the 1820s not only did the calorie supply increase rapidly, but the amount of animal food also increased, although relatively slowly (table 3.2); as late as the 1870s 70 per cent of French calories were still derived from bread and potatoes.[10]

At the end of the nineteenth century there began a fundamental change in the diet of Western Europe. Once national food supplies had reached 3000 calories per caput there was comparatively little further increase in total calorie supply, or for that matter in total protein supply. Instead there were major changes in the composition of the diet. Animal foods began to provide an increasing proportion of calorie intake. Thus in France animal food provided one-fifth of all calories in 1900, but this had risen by the 1960s to one-third. In 1900 75 per cent of all protein intake was derived from vegetable foods, but by the 1960s two-thirds of all protein came from animal foods. Not only have vegetable foods provided a declining percentage of calorific intake this century, but the absolute quantities of bread and potatoes consumed has declined. There has also been an increase in fat and sugar consumption.[11]

Since 1800 there have been fundamental changes in food supplies in Western Europe. Paul Bairoch has calculated, using FAO estimates of nutritional requirements and the age and sex structures of the French population in 1830, that the minimum national requirements were 2300 calories per caput per day. In the early nineteenth century France, Norway and Germany all had supplies below this level. Hence, even if national supplies had been equitably distributed, they would have been

Table 3.3 Income and nutritional value of household budgets, England, 1887–1901

	Income class (shillings per week)	Calories per caput per day	Grams of protein per caput per day
A	Under 18	1578	42
B	18–21	1964	51
C	21–30	2113	58
D	30 and over	2537	72
E	Families with servants	3256	96

Source: D. J. Oddy, 'Working class diets in late nineteenth-century Britain', *Economic History Review*, **23**, 1970, pp. 314–23

insufficient. From the 1820s there were rapid increases in national food supplies, such that by the mid nineteenth century most countries had available food supplies sufficient to eliminate hunger. The increase in calorie supplies came mainly from increases in supplies of bread and potatoes. But from the 1890s there was a steady increase in the proportion of food supplies derived from animal foods.[12]

Hence at various times in the nineteenth century the national food supplies of most West European countries reached a level which was, probably for the first time, sufficient to eliminate undernutrition and malnutrition if distributed according to need. But as has already been seen, national food supplies are not a sufficient guide to the extent of hunger.

Household consumption

That the food supplies of West European countries were not equally distributed in the nineteenth century, in the first half of this century, or indeed today, is undoubted; nor is there any doubt that low incomes were the prime cause of low food intake. D. J. Oddy has shown, from a number of household budgets of the late nineteenth century, how calorific intake varied with income in England (table 3.3).

There is little doubt that elsewhere in Europe the nutritional intake of poor families was inadequate in the later nineteenth century. Budgets of poor households in France collected by Le Play and his followers between 1856 and 1901 gave a mean of 2791 calories, but the range was from 1638 to 4678. One peasant household had a daily supply of 4161 calories, 86 per cent derived from bread.[13]

Most historians seem agreed that, although national food supplies were sufficient for much of the second half of the nineteenth century in Britain, there must have been widespread malnutrition because of the very marked inequalities in income distribution and the low level of wages of the poorest, although there is no reliable way of measuring the extent of malnutrition or (less common) undernutrition.[14] The Edwardian period saw considerable concern about the extent of malnutrition. The health and nutrition of the working class became a matter of government enquiry, prompted initially by the discovery that many volunteers for the Boer War were rejected on medical grounds, and that part of the cause of this was malnutrition. The reports of the school inspectors also revealed much evidence of malnutrition among children. On the eve of the First World War the Board of Education stated that 11 per cent of English children had defective nutrition.[15]

Malnutrition persisted into the inter-war period. By then more was known of the causes of malnutrition, more data were available on nutrition and more scientific surveys could be carried out. The depression of the 1930s and the large number of unemployed drew attention to this problem. Sir John Boyd Orr estimated that half the population of Britain had incomes too small to provide a diet sufficient for optimum growth and health. The League of Nations was much concerned with the extent of malnutrition in Europe, and had laid down minimum nutritional requirements for health. The British Medical Association calculated that one-third of the population of Britain lacked the incomes to acquire this diet. Nor was malnutrition confined to Britain. A later League of Nations report stated that there was no country in Europe where all the population reached the minimum diet and that scurvy, anaemia and rickets were widespread. Government reports in the United States also revealed an alarming level of malnutrition.[16]

Changes in health, nutrition and mortality

Modern inquiries into undernutrition in the developing countries have compared the growth of children in height and body weight with a healthy norm to obtain some measure of the extent of hunger. Long-term changes in the height of children, adolescents and young men can also indicate how the nutritional status of a population has changed over time, although of course other factors may influence height. Similarly, the age of menarche in girls is influenced by a variety of factors, but long-term changes are thought to reflect changes in nutrition.

There is evidence to suggest that there have been significant increases in average height in Europe over the last 100 years and probably over even longer periods. Thus adult males in Europe are now 6–9 cm taller than they were in the 1870s. In Norway, for example, although there was little increase in the average height of adults between 1760 and 1830, it rose by 3 mm per decade over the next forty-five years and by 6 mm per decade from 1875 to 1935. Records of conscripts in Holland show a similar rate of increase, although height actually fell between 1820 and 1860. Studies of children also show that the average height for particular ages has risen. Thus the height and weight of school-children in Oslo rose between 1920 and 1940, fell during the war, and increased thereafter. In Sweden the average height of 18-year-old boys rose from 168.5 cm in 1883 to 178.4 cm in 1969–71. Rather more dramatic evidence is available on children in London. The records of the Marine Society, who placed poor children in the navy, suggest that the average height of boys born after 1815 was 7.5–12.5 cm greater than those born in the 1770s. Thirteen year olds born 1753–80 averaged 130 cm; a sample of 13-year-old boys in the London County Council area in 1965 averaged 155.5 cm.[17]

Differences in nutrition are also a major factor in the age of menarche among girls. At any one time there are differences according to income, those in richer families reaching menarche earlier than those in poorer. There are also changes over time; a study of working class women in Norway suggests that the average age of menarche fell from 15 years in the 1840s to 13 in 1950. Comparable declines have been recorded in England.[18]

Although nutrition may not be the only cause of changes in height, body weight and menarche over time – genetic factors may help explain long-term increases in height – it is undoubtedly important, and these changes reflect improved nutrition in Europe over the last 150 years. It is noteworthy that recent studies suggest that native born whites in the United States were not far short of modern heights as early as the late eighteenth century and were taller than Europeans at the time. The difference in height between adult males in Europe and those in the United States has markedly diminished in this century, as have differences in average height between children and adults of different income groups within developed countries. It is important to note that the increase in height – and so of nutrition – has not been continuous in the United States. Heights increased until the 1830s, and then fell, recovering only after 1890; this has been attributed to a fall in calorific intake as agricultural output failed to keep up with population growth. A similar decline took place in Sweden in the first part of the nineteenth century.[19]

Child nutrition and mortality

Modern nutritional experts believe that protein calorie deficiency is a major cause of child mortality, i.e. the death rate of children aged 1–5.

However, the close interrelationship between infection and malnutrition makes authoritative pronouncements on this difficult, and certainly a decline in the child mortality rate may reflect improvements in sanitation and water supplies as much as improvements in nutritional status.

There is little evidence of *acute* protein calorie malnutrition in Britain or the United States in the nineteenth century, although symptoms that appear to resemble kwashiorkor were described in Germany, Switzerland and Italy. Moderate and mild forms, however, are thought to have been widespread. In those countries for which records of child mortality exist there was a decline in the nineteenth century that has continued to the present; in most countries the decline in infant and child mortality preceded that of the older age groups.[20]

The longest available series of child mortality rates is for Sweden. In the late eighteenth century rates exceeded forty per thousand but began a continuous decline from the second decade of the nineteenth century, reaching twenty-five per thousand in the 1830s, sixteen per thousand in 1899–1902 and 4.1 per thousand in 1929–32. The rate now stands at 0.4 per thousand. The decline in Denmark followed a very similar course. In England the rate was higher than in Sweden in the mid nineteenth century – at about thirty-five per thousand – and did not begin a permanent decline until the 1860s; it remained above the Scandinavian rates until after the Second World War. France and the Netherlands also showed improvement in expectation of life at younger ages from the mid nineteenth century.[21] There remains considerable controversy about the evidence of child mortality rates as a guide to improved nutritional status. Some authorities believe that falling mortality in the nineteenth century was due to better food supplies, a reduced rate of malnutrition and hence greater resistance to infectious disease. Others have doubted the significance of nutrition and have attributed most of the decline to improvements in public health, noting that the life expectancy of well-fed American families in the mid nineteenth century was not very different from the supposedly malnourished West Europeans. It can be said, however, that in the early nineteenth century child mortality rates were comparable with those in the poorer developing countries in the 1940s – over forty per thousand – and have fallen to levels below one per thousand today.[22]

The combined evidence of national food supplies, household budgets, heights and falling child mortality rates suggests great improvement in the nutritional status of the population of Western Europe from the eighteenth century to the present. In the late eighteenth century and early nineteenth century national food supplies were as low – indeed lower – as those at present in the poorer parts of Africa and South Asia. The nineteenth century saw a rapid growth in calorie supplies, this century the growing importance of animal foods. But malnutrition

persisted until the 1930s because significant proportions of the population had incomes too low to provide an adequate diet. By the 1930s
Portugal was the only country in Europe with a national food supply
below national requirements.

THE DEVELOPING COUNTRIES

Little is known of the nutritional status of the populations of Africa,
Asia and Latin America before the twentieth century. The first estimates of calorie supplies were made by the FAO in the first *World Food
Survey*, published in 1946 but relying upon data collected in the 1930s.
It is clear that the great difference between the developing and developed countries already existed (figure 3.3) although calorie availability
per caput was lower than at present in both regions. Clearly the comparatively low supplies of developing countries today cannot be due, as
is sometimes suggested, solely to the rapid population growth of the last
forty years.

If calories per caput were inadequate in the 1930s in much of Africa,
Asia and Latin America, so too was the quality of the diet. There are
no national records of protein per caput in the 1930s. However, M. K.
Bennett estimated the proportion of total calorie intake derived from
cereal and root crops, and argued that if this exceeded 80 per cent the
population would be malnourished.[23] The whole of Asia – except Japan
– and most of Africa fell into this category, whereas parts of Eastern
Europe and North Africa had between 70 and 80 per cent, comparable
with France in the early nineteenth century (figure 3.4). Nowhere in
Latin America, however, did the proportion exceed 70 per cent. North
America, Australasia and North West Europe had less than 40 per cent
of their calories from cereals and roots. By the 1980s (figure 3.5) there
had been marked changes in both the developed and the developing
countries; in only a handful of countries did the proportion still exceed
80 per cent, and most of the population of Latin America lived in
countries where the proportion was below 50 per cent. The difference
within the developed region between Western Europe on the one hand
and southern Europe, Eastern Europe and the Soviet Union on the
other still persists, although the proportion of calories derived from
cereals and roots has declined in the latter areas.

Since the end of the Second World War more detailed and more
reliable evidence on national food supplies has become available. Furthermore the FAO of the United Nations has published five world food
surveys. These should give some indication of the trend in the numbers
malnourished in the developing world (table 3.4). The first *World Food
Survey* (1946) dealt with the 1930s and estimated that then one-half of

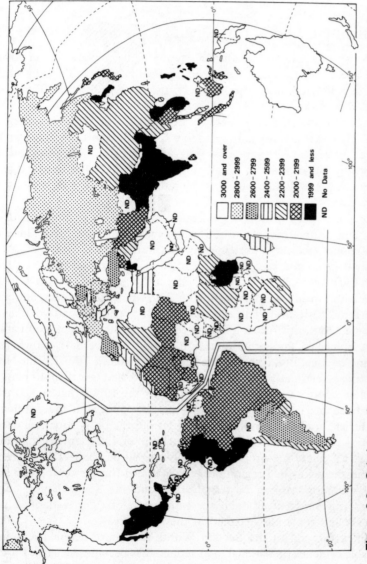

Figure 3.3 Calorie supply per day per caput, 1934–8 *Source:* FAO, *World Food Survey*, Washington, D.C., 1946

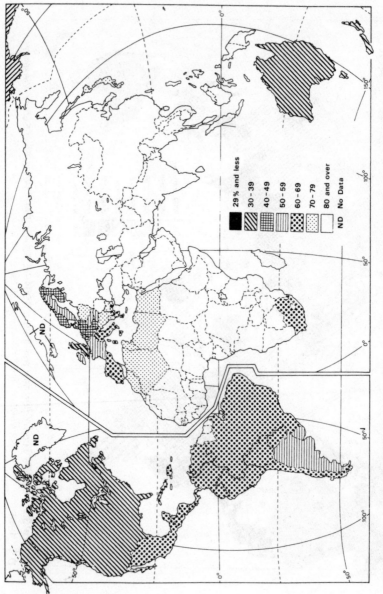

Figure 3.4 Percentage of total calorie supply per caput per day derived from cereals and roots, 1934–8 *Source*: M. K. Bennett, 'International contrasts in food consumption', *Geographical Review*, 31, 1941, pp. 365–76

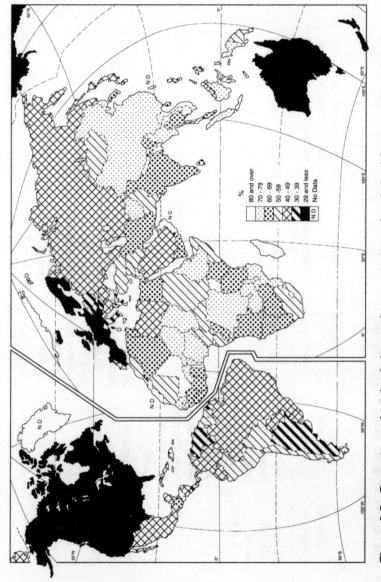

%
80 and over
70 - 79
60 - 69
50 - 59
40 - 49
30 - 39
29 and less
N D   No Data

Figure 3.5 Percentage of total calorie supply per caput per day derived from cereals and roots, 1986–8 *Source*: FAO, *Food Balance Sheets 1986–88*, Rome, 1990

Table 3.4 FAO estimates of the extent of world hunger

| Date of survey | Size of sample | Extent of hunger | Number or proportion | | Total number undernourished or malnourished (millions) | As a percentage of world population | World population (millions) |
			Undernourished	Malnourished			
1934–8	70 countries with 90% of world population	'Half the world's population ...live ...at a level of food consumption which was not high enough to maintain health'			1050	50	2100
1934–8	80% of the world's population	Not stated	38.6% of the population lived in countries with daily food supplies less than 2200 calories	59% of the population lived in countries where daily supply of animal protein is less than 15 grams	1240	59	2100

1949–50	80% of the world's population	Not stated	59.5% of the population lived in countries with daily food supplies less than 2200 calories	58% of the population lived in countries where daily supply of animal protein is less than 15 grams	1490	59.9	2500
1957–9	80 countries with 95% of world population	10%–15% of the world's population are undernourished and up to half suffer from hunger or malnutrition	10%–15% of the world population	'The incidence malnutrition in the less developed countries is estimated at 60%'	'up to half = 1430' 'the incidence of malnutrition = 1160'	40	2870
1972–4	58 less developed countries containing 56% of the population; Asian Communist countries excluded	455 million have a food intake below 1.2 BMR	455		455	11.8	3830

(cont.)

Table 3.4 (Cont.)

| Date of survey | Size of sample | Extent of hunger | Number or proportion | | Total number undernourished or malnourished (millions) | As a percentage of world population | World population (millions) |
			Undernourished	Malnourished			
1979–81	112 less developed countries, excluding Asian Communist countries	335 million have below 1.2 BMR, 494 million have below 1.4 BMR	355–494 million, 15%–23% of population of developing countries		335–494	7.5–11.1	4450

BMR, basal metabolic rate.
Sources: FAO, *World Food Survey*, Washington, D.C., 1946; *The Second World Food Survey*, Rome, 1952; *The Third World Food Survey*, Rome, 1963; *The Fourth World Food Survey*, Rome, 1977; *The Fifth World Food Survey*, Rome, 1987

the world's population had inadequate diets. *The Second World Food Survey* (1952) made no estimate of the extent of hunger, but noted that 59.5 per cent of the population lived in countries where daily food supplies were less than 2200 calories, and a similar proportion lived in countries where daily supplies of animal protein were less than 15 grams; it must be assumed that it is this information that Sir John Boyd Orr used to make his statement that two-thirds of the world's population lived lives of malnutrition and hunger.[24] In *The Third World Food Survey* (1963) it was argued that only 10–15 per cent of the world's population had an insufficient calorie supply, but that 60 per cent of the population of the developing countries suffered from malnutrition caused by a lack of animal protein. By the time of *The Fourth World Food Survey*, it had been decided that animal protein was not essential for health and that a diet which provided a sufficient supply of calories would also provide sufficient protein and vitamins. This survey found that only 455 million in the non-Communist developing countries suffered from undernutrition. Estimates for the Communist countries would make the total 630 million, about 15 per cent of the world population. The most recent survey found that in 1979–81 only 335 million received less then 1.2 BMR, a further decline in numbers and proportion.

The apparent decline in the proportion and the absolute numbers suffering from undernutrition and malnutrition in the FAO surveys is a result of the very different ways in which each survey defined and measured hunger; hence the fall from 1950 to 1979–81 cannot be used to show that the problem of hunger has diminished. It is, however, possible to use the figures on available food supplies published by the FAO, if not to provide an accurate estimate of the extent of hunger, at least to indicate some trends between 1950 and 1988. There are four ways of doing this.

Trends in available food supplies per caput   Between 1950 and 1988 available food supply per caput rose in the world as a whole, in the developed world, in the developing world and in all the major regions (table 3.5). There is no information on how these supplies were distributed among the population in either year, but in 1950 available supplies in Africa and Asia were very low, and even if supplies had been equally distributed would probably not have provided an adequate diet. However, it is very clear that the dramatic decline in food supplies per caput predicted by many in 1950 has not occurred.

Numbers living in countries with less than 2200 calories per caput per day   In *The Second World Food Survey* attention was drawn to the numbers living in countries with less than 2200 calories, the assumption being that a

Table 3.5   Average available food supply per caput per day, 1950–88

	1950	1961–5	1969–71	1978–80	1987–9	1950–88, percentage change
Developed						
Europe	2689[a]	3420	3339	3477	3459	28.6
North America	3131	3492	3467	3624	3656	16.8
Oceania[b]	3176	3432	3360	3257	3220	1.4
USSR	3020	3542	3388	3486	3380	11.9
Total, developed	2878	3471	3382	3486	3415	18.7
Developing						
Asia[c]	1924	2068	2192	2326	2487	29.3
Africa[d]	2020[e]	2165	2276	2311	2360	16.8
Latin America	2376	2413	2531	2591	2724	14.6
Total, developing	1977	2115	2239	2350	2474	25.1
World	2253	2494	2537	2617	2703	20.0

[a] No data for Eastern Europe: assumed to be the same as Western Europe for aggregated figures.
[b] Australia and New Zealand only.
[c] Includes China, Japan and Israel.
[d] All Africa, including South Africa.
[e] Data for 44 per cent of population only; other countries are assumed to have the same average.
Sources: FAO, *The Second World Food Survey*, Rome, 1952; *Production Yearbook 1976*, vol. 30, Rome, 1977; *Production Yearbook 1982*, vol. 36, Rome, 1983; *Production Yearbook 1989*, vol. 43, Rome, 1990

large proportion of the population would be undernourished. This seems reasonable, for when in *The Fourth World Food Survey* national minimum requirements were estimated for each developing country, in most cases they were more than 2200 calories. There was little change in the proportion of the world's population living in such countries between 1950 and 1961–5, while the absolute numbers greatly increased (table 3.6). However, since 1961–5 both the proportion and the absolute numbers have declined, and by 1984–6 only one-tenth of the world's population lived in countries with available food supplies of less than 2200 calories per day.

Table 3.6 World population in three intake classes (calories per caput per day), 1950, 1961–5, 1978–80 and 1987–9

Intake class	Percentage of total population					Numbers (millions)				
	1950	1961–5	1978–80	1987–9		1950	1961–5	1978–80	1987–9
Less than 2200	59.8	57.6	25.6	27.3		1251.1	1822.7	1103.5	1348.9
2200–2599	9.7	11.5	38.8	9.5		205.7	365.1	1665.4	467.7
2600 and over	30.5	30.9	35.6	63.2		633.1	979.2	1532.0	3107.8
Total	100.0	100.0	100.0	100.0		2089.9	3167.0	4300.9	4924.4

Sources: FAO, *The Second World Food Survey*, Rome, 1952; *Production Yearbook 1976*, vol. 30, Rome, 1977; *Production Yearbook 1982*, vol. 36, Rome, 1983; *Production Yearbook 1957*, vol. 8, Rome, 1958; *The State of Food and Agriculture 1989*, Rome, 1989; *Production Yearbook 1988*, vol. 42, Rome, 1989; United Nations, *Demographic Yearbook 1950*, New York, 1951

Table 3.7   Population receiving less than the minimum requirement of
1.2 BMR, 1950–88

	1948–50	1961–5	1972–4	1978–80	1986–8
Numbers (millions)					
Africa	60	94	83	72	181
Latin America	46	54	46	41	36
Asia (including China)	444	502	397	421	477
Total	550	650	526	534	694
Percentage of population					
Developing countries	34	29	20	17	18
All countries	23	21	13	12	14

Sources: FAO, *The Fourth World Food Survey*, Rome, 1977; *The State of Food and Agriculture, 1981*, Rome, 1982; *Production Yearbooks*, Rome, 1948 onwards

Changes in the numbers receiving less than 1.2 BMR   In *The Fourth World Food Survey* FAO statisticians calculated the numbers receiving less than 1.2 BMR in 1972–4. A simplified way of calculating these numbers can be applied to FAO food availability statistics for other years (table 3.7). The numbers receiving less increased from 1950 to 1961–5, declined in the 1970s, but has increased again since 1980. The proportion receiving below the critical intake fell continuously from 23 per cent in 1950 to 12 per cent in 1980, but has since risen slightly (table 3.7). In *The Fifth World Food Survey* a different methodology was used by the FAO to calculate the numbers receiving less than 1.2 BMR (table 2.10); although not covering the whole of the post-war period, these figures do suggest that the downward trend in the proportion undernourished and the relatively small increase in the numbers in the 1970s was arrested in the 1980s, a period of severe economic crisis in Africa and Latin America.

Changes in the number of countries not meeting national minimum requirements   In the early 1960s and 1970s nearly all the countries in Asia and Africa had calorific supplies insufficient to meet national requirements as defined in *The Fourth World Food Survey*, as had the majority of countries in Latin America (figure 3.6). By the 1980s, however, the oil exporters of North Africa and the Middle East were importing enough food to provide supplies above this minimum, and China by a combination of improved production and imports had supplies above the minimum. Over the last thirty years the problem of insufficient

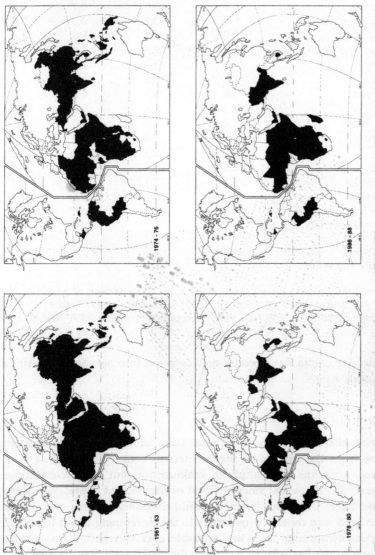

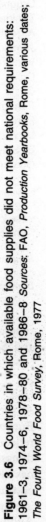

Figure 3.6 Countries in which available food supplies did not meet national requirements: 1961–3, 1974–6, 1978–80 and 1986–8 *Sources*: FAO, *Production Yearbooks*, Rome, various dates; *The Fourth World Food Survey*, Rome, 1977

Table 3.8 Mortality rate in children aged 1–4 years (per 1000 children)

	1950–2	1960–2	1966	1975a
Sweden	1.3[b]	0.9	0.7[c]	0.4
United Kingdom	1.4[b]	0.9	0.8[c]	0.6[d]
Uruguay	2.2	1.3	1.3	1.6
Trinidad and Tobago	5.8	2.5	2.0	1.9
Argentina	5.0	4.3	2.4	0.7
Jamaica	10.5	6.8	4.7	–
Venezuela	11.9	5.7	4.9	1.6
Chile	12.9	8.2	5.0	2.2
Costa Rica	14.6	7.5	6.0	3.1
Panama	9.5	7.9	8.0	3.4
Peru	19.8	15.7	10.5	8.3
Mexico	28.6	13.8	10.9	–
El Salvador	31.1	17.1	13.5	8.8[e]
Guatemala	46.3	32.4	29.5	–

[a] Male child mortality only.
[b] 1950.
[c] 1965.
[d] England and Wales.
[e] 1971.
Sources: J. M. Bengoa, 'Recent trends in the public health aspects of protein–calorie malnutrition', WHO Chronicle, 24, 1970, pp. 552–61; United Nations, Demographic Yearbook 1979, New York; Demographic Yearbook 1986, New York

supplies per capita has become concentrated into sub-Saharan Africa and South Asia.

Declines in child mortality

As seen earlier there was a continuous fall in child mortality in Western Europe from the middle of the nineteenth century. This must have reflected a decline not simply in malnutrition but also in infectious diseases.[25] In the period since the end of the Second World War there has also been a marked decline in child mortality – those over 1 year and under 5 – in those few countries in the developing world for which age specific mortality rates are available over this period (table 3.8). Although in the few African countries for which estimates are available, such as Togo and Benin, child mortality is over forty per thousand (comparable with Sweden in the late eighteenth century), in many developing countries it has fallen below ten per thousand, a rate not achieved in England and Wales until 1920.[26]

Table 3.9 Changes in the probability of dying between birth and age 5 per 1000 live births, median for regions

	1960–5	1965–70	1970–5	1975–80	1980–5
Latin America	105	100	89	52	42
Asia	128	114	91	68	55
Middle East	195	183	180	130	141
Africa	233	226	220	193	198

Sources: K. Hill and A. R. Pebley, 'Child mortality in the developing world', Population and Development Review, **15** (4), 1989, pp. 657–87

Data on the mortality of all children under 5 since the 1960s shows a continuous fall until the late 1970s. In both Africa and the Middle East the probability of dying before the age of 5 then increased slightly (table 3.9) in the 1980s, possibly as a result of the economic recession of the later 1970s and due to cuts in health services in Africa in the 1980s. However, there is little doubt that the general trend in child mortality in the developing countries has been downwards since 1950.[27]

CONCLUSIONS

Although it is not possible to make any firm statements about the extent of hunger over the last 200 years, some conclusions can be drawn.

First, in the early nineteenth century many countries in Western Europe had available food supplies which were below the national minimum requirements, and from this it can be inferred that a considerable but unknown proportion of the population was both undernourished and malnourished. This was due to the very low productivity of agriculture and the poverty of much of the population.

Second, by the later nineteenth century most countries in Western Europe had national food supplies that would have provided an adequate diet for everyone if distributed according to needs. But contemporary descriptions suggest that malnutrition was still widespread. The evidence of contemporary working class household budgets, the reports of school inspectors and the rejection rates of army volunteers all suggest widespread malnutrition in England.[28] Malnutrition was still widespread in the 1930s, not only in Britain but also in Europe and the United States. Indeed it was not until the period since the end of the Second World War that malnutrition declined to a very small proportion of the population of the developed world. The principal reason for the earlier widespread malnutrition was poverty; substantial proportions of the population had incomes too small to buy an adequate diet. Thus it was not until the remarkable economic growth in the 1950s and 1960s that the incomes of the least well paid rose above the levels necessary to provide an adequate diet. There were other factors that contributed to the decline of malnutrition. The first was the improved knowledge of nutrition; it was not until the early part of this century that the significance of vitamins was known. The second was the decline in infectious disease from the late nineteenth century; as cleaner water supplies and more effective sanitation were provided, so the incidence of gastric diseases declined, and this may have reduced the influence of malnutrition. Third, in much of Europe and the United States legislation to prevent the sale of diseased and adulterated food must have not only reduced poisoning but also increased the nutritional value of foods sold in shops. Finally, in some countries the state made efforts to improve the diet of children; thus, for example, in Britain the provision of school meals and milk may have helped reduce the extent of malnutrition.

Thus it is clear that the provision of adequate national food supplies is not sufficient to ensure the elimination of undernutrition and malnutrition. The poor must have enough money to purchase the minimum diet. It is disturbing to note that there was a lag of up to a century between some European countries' achievement of an adequate national food supply and the elimination of malnutrition. But equally clearly in the early nineteenth century much of Europe had insufficient food supplies for its population quite regardless of income differences; this was attributable to the very low level of agricultural productivity, the high cost of transporting food, the difficulties of preserving perishable foods and the losses of foods in storage. Nor are all countries in the developing world today capable of providing an adequate food supply even when imports are taken into account (see figure 2.4).

In the post-war period malnutrition has been largely confined to the developing countries; paradoxically, at the time it became fashionable to speak of the *world* food problem, malnutrition became a problem almost exclusively of Africa, Asia and Latin America. It would seem likely that between 1945 and 1980 the proportion of the population of the developing countries suffering from undernutrition declined; even with the great increase in population, the absolute numbers did not greatly increase. But in the 1980s this long-term decline suffered a set back, and both absolute numbers and the proportion undernourished increased.

4

Population and Poverty

In the 1930s malnutrition was thought to be due to poverty; few attributed the problems in Europe or the United States to excessive population growth. Indeed the prime concern of demographers at the time was why the rate of European population growth was so low. The *World Food Survey* (1946) also argued that the extent of hunger was due to poverty and had little to say about population growth, but by the time of *The Second World Food Survey* (1952) there were fears that the rapidity of population growth in the developing countries, which was becoming apparent as the first censuses taken for some time became available, might prevent the elimination of hunger.[1] Throughout the 1950s and 1960s it came to be assumed that the major cause of hunger in the developing countries was population growth: quite simply the numbers in Africa, Asia and Latin America were growing faster than the capacity of agriculture to supply food.

By the 1970s, however, this point of view was under challenge. It could be shown first that per caput food supplies were very low in the developing countries in the 1930s (see figure 3.3), so that although post-war population growth may have prevented the improvement of post-war consumption levels, it could hardly be said to have caused their lowness. Second, after all the gloomy predictions of the 1950s and 1960s, it seemed clear that at the world level food production had kept up with population growth, and that world food supplies per caput were somewhat above the level of the 1930s or early 1950s. Consequently it became common to attribute hunger in the developing countries to poverty, and efforts were made to measure the numbers without the resources to acquire enough food.

Unfortunately these conflicting views of the problem have been connected with opposing ideological views of the world. There has been a tendency for those who believe that inequality of income distribution is the only cause of hunger to ignore the consequences of population growth, and for those who believe that poverty is solely a result of too rapid population growth to neglect income inequalities. There seems no need to take such entrenched positions. At any one time the existence of undernutrition or malnutrition is surely due to poverty – to the lack of income, employment opportunities or sufficient land. But, over time, population growth is equally surely one possible cause of poverty.

The aim of this chapter is to outline the growth of population in the world since 1950, with some reflections on earlier growth, as a background for the discussion of the growth of food output per caput in chapter 5. Some of the main features of poverty are then discussed.

THE POST-WAR GROWTH OF POPULATION

The rapid growth of population since 1950 has increased the number of consumers of food, and the changing geographical distribution has influenced levels of consumption. Increasing numbers have also, of course, increased the number of producers of food. Both aspects must be noted. There is no doubting the rapidity of population growth in the developing countries between 1950 and 1988; not only did the population of Africa, Asia and Latin America increase far more rapidly than the developed countries in the same period, but far more rapidly than in any part of the world before 1950.[2]

Between 1950 and 1990 the population of the developing world more than doubled (table 4.1), and in every quinquennia since 1950 the rate of increase has been over 2.0 per cent per annum (table 4.2). There are, however, important differences between the three major developing regions. In Latin America the rate of increase reached a peak in the 1960s and has since fallen, whilst in Asia a peak was reached in the same decade and has since declined, falling below 2 per cent per annum in the late 1970s and early 1980s. In Africa, however, the rate of increase rose in every quinquennia from the 1950s to the 1980s.

The population of the developed world has only increased by 44 per cent in the same period and rates of increase, except during the period of substantial emigration to North America and Australasia after the Second World War, have never matched those found in the developing world. Further the rate of increase in the developed countries has been in continuous decline since the early 1950s (table 4.2).

There are, however, quite significant regional differences within the two spheres. Within the developed countries the increase for 1950–88 has been least in Western Europe and most in North America, the

Table 4.1   Population, major regions, 1950–90 (millions)

	1950	1960	1970	1980	1990	Increase, 1950–90	Percentage increase, 1950–90
Developing							
Africa	224	281	363	481	642	418	186
Latin America	165	218	285	362	448	283	172
Asia	1375	1667	2101	2583	3112	1737	126
Total, developing	1764	2166	2749	3426	4202	2438	· 138
Developed							
Europe	393	425	460	484	500	107	27
North America	166	199	226	252	275	109	66
USSR	180	214	243	265	288	108	60
Australasia	11	14	16	18	20	9	82
Total, developed	750	852	945	1019	1083	333	44
World	2514	3018	3694	4445	5221	2707	107

Source: United Nations, *Demographic Yearbook 1988,* New York, 1990, p. 161

USSR and Australia. Within the developing world the increase has been greatest in central America, northern South America, parts of Africa, South-west Asia and parts of South-east Asia. However, in China and India, by far the most populous countries in the world, the post-war population increase has been comparatively modest; indeed parts of the developed world have increased as rapidly. Thus China's population rose by 72 per cent between 1950 and 1980, India's by 85 per cent. In the same period, the population of Canada rose by 74 per cent. The difference lies of course in the absolute numbers. India's population in 1980 was 305 million more than in 1950, China's 400 million and Canada's only 10 million.[3]

TRENDS IN MORTALITY AND FERTILITY

Although some parts of the developing world experienced an increase in fertility in the 1950s and 1960s, the main cause of the great upsurge in population has been a decline in mortality combined with, until recently, little or no change in fertility. In most developing countries the crude death rate is now half what is was just after the Second World

Table 4.2 World and major regions: rates of population increase, 1950–85 (average rate of increase in per cent per annum)

	1950–4	1955–9	1960–4	1965–9	1970–4	1975–80	1980–5
Developing							
Africa	2.18	2.35	2.48	2.63	2.69	2.95	2.95
Latin America	2.74	2.76	2.80	2.60	2.48	2.28	2.19
Asia	1.9	1.95	2.19	2.44	2.27	1.86	1.86
Total, developing	2.05	2.14	2.35	2.54	2.39	2.10	2.10
Developed							
North America	1.80	1.78	1.49	1.13	1.06	1.06	1.00
Europe	0.79	0.8	0.91	0.67	0.58	0.45	0.32
USSR	1.71	1.77	1.49	1.01	0.96	0.82	0.84
Australia and New Zealand	2.33	2.18	1.99	1.85	1.67	1.27	1.30
Total, developed	1.28	1.25	1.19	0.91	0.86	0.73	0.65
World	1.8	1.86	1.99	2.06	1.97	1.74	1.74

Source: United Nations, *World Population Prospects 1988*, Population Studies No. 106, New York, 1989

Table 4.3 Crude birth and death rates, 1950–85

	Crude birth rates (per thousand)		Crude death rates (per thousand)	
	1950–5	1980–5	1950–5	1980–5
Africa	48.9	45.5	27.0	16.4
Asia	42.9	28.4	24.1	9.8
Latin America	42.5	30.9	15.3	8.0
Europe	19.8	13.4	11.0	10.5
Oceania	27.6	20.6	12.4	8.1
North America	24.6	15.6	9.4	8.5
USSR	26.3	19.1	9.2	10.7
World	37.4	27.7	19.7	10.4

Source: United Nations, *World population prospects 1988*, Population Studies No. 106, New York, 1989, pp. 44–7

War (table 4.3), and there have also been dramatic declines in infant and child mortality. The reasons for this decline are not entirely clear; a number of explanations have been put forward. First the improvements in chemical therapy in the 1930s and 1940s have been extended from the developed to the developing countries. Antibiotics have cured, and vaccines have increased immunity, to a variety of diseases. Second, advances in preventive health have reduced the prevalence of certain contagious diseases. The improvement of water supplies and sewage disposal has reduced the incidence of gastrinal diseases, and immediately after the Second World War the spraying of badly drained areas with DDT destroyed mosquitoes and thus reduced malaria. However, malaria has reappeared as a major disease in recent years as mosquitoes have developed immunity to DDT and its successors. Third, the increase in income per caput in the less developed countries may also have contributed to the decline in the death rate by improving nutrition, increasing the provision of medical services and raising standards of literacy and general education. The better education of women has been particularly important in the reduction of infant and child mortality. Although crude death rates are still declining in the developing world, in most countries the rate of decline was greater before the mid 1960s than since.[4]

The decline in mortality that was so rapid in the 1950s was not at first accompanied by any change in the crude birth rate, which was high, particularly in Africa and Latin America. This was because marriage was almost universal in the developing countries, most women married when very young, and there was little attempt to control births within marriage. However, by the mid 1960s signs of a decline in fertility were becoming apparent in countries with a Chinese population, such as Taiwan, Malaya and Singapore, and also in China itself. Later there was a decline in other parts of Asia and in Latin America, but not in Africa (table 4.3). This fall in fertility seems to have been due not only to the spread of family planning methods, but also to a rise in the age of marriage of women and some decline in the proportion of women getting married.[5] The most dramatic reduction in fertility appears to have been achieved in China, where since the early 1960s the government has encouraged the adoption of birth control methods, discouraged early marriage and penalized the parents of large families. Since 1979 there has been a campaign to encourage one-child families. The crude birth rate is thought to have fallen from thirty-three per thousand in 1970 to seventeen per thousand in 1977. But although there has been a substantial decline in crude birth rates in the developing world, and although mortality has been falling more slowly since 1970, the rate of natural increase continues to be high; between the 1960s and the 1980s some reduction of the rate of increase had been achieved in Latin America and Asia but not in Africa (table 4.2), where rates of population increase are currently higher than at any time in the past.[6]

Population growth in the developed countries merits less attention for there are now comparatively few problems of undernutrition or malnutrition. Yet the population has increased since 1950; there were 333 million more people to be fed in 1990 than there were in 1950. This is small compared with the 2438 million people added in the developing countries, but none the less has required an increase in food output (table 4.1). In 1950 crude death rates were already low in the developed countries, having been declining for at least a century; although the mortality rates of most age groups have fallen further since 1950, the ageing of the population has ensured that the crude death rate has not been greatly reduced (table 4.3). Thus in most parts of the developed world it has been trends in fertility that have determined the rate of population increase since 1950. By the 1930s fertility had fallen very low in most of the Western world, except in the Soviet Union or Eastern Europe. There was a rise in fertility after the end of the Second World War in Western Europe and the European settlements overseas, but this ended in the mid 1960s, and by the 1970s fertility had fallen so low that annual deaths were exceeding births in some West European states.[7]

POPULATION GROWTH IN A LONGER PERSPECTIVE

In the 1950s much emphasis was put upon the dramatic decline in the death rate in the developing countries. Some historians had contrasted the rapid growth of European populations in the nineteenth and twentieth centuries with the apparently slow increases in Africa, Asia and Latin America. It was assumed that economic growth in the developed world had reduced mortality in Europe and North America, but that there was little progress in mortality reduction in the developing countries until after 1945. The population explosion in the less developed countries was thought to have no precedent.

This is not now thought to be so. Although the current rates of increase of 2 per cent per annum or more in the developing world are higher than those found before 1950, rates of increase were then already comparatively high. Paul Bairoch believes that the developing countries were increasing at 0.5 per cent per annum in the first two decades of this century, then at over 1 per cent in each decade until 1950 and since then at over 2 per cent.[8] Reliable information on population increase in the developing world is hard to come by before the 1940s, but certainly many countries appear to have experienced quite rapid increases in the first half of the twentieth century. Egypt's population increased more than 1 per cent per annum in every decade after 1900, as did Brazil's, and Thailand's population was increasing at over 2 per cent per annum in both the 1920s and the 1930s. The reason for this brisk increase in population was apparently a decline in mortality, although this is well documented only for Latin America, where mortality has been declining since 1900 and probably before then. Thus expectation of life at birth in Brazil rose from 29 years in 1900 to 45 years in 1950, and in Mexico from 25 years to 47 years.[9]

Indeed over the last 200 years the rate of population increase in the countries now defined as developed and developing has not greatly differed (table 4.4). The upturn in Europe's population, which occurred in the eighteenth century, was once thought to be unique. But modern estimates of population elsewhere in the world suggest that the increase was almost universal.[10] The difference between the rate of population growth in the developed and the developing countries was most marked between 1850 and 1899 (table 4.4). However, the population of the developing countries was heavily weighted by that of China, whose population fell during the Taiping rebellion and recovered only slowly thereafter. In contrast in some parts of the developing world population increase was very rapid. Java's population tripled

Table 4.4 Population growth in the developed and developing worlds, 1750–1980

Average rate of increase (per cent per annum)

	1750–99	1800–49	1850–99	1900–49	1950–80	1750–1980
Developed[a]	0.4	0.7	1.0	0.8	1.0	0.75
Developing	0.4	0.5	0.3	0.9	2.3	0.7

Percentage of world population in developed and developing countries

	1750	1800	1850	1900	1950	1990
Developing	75	75	73	66	67	76
Developed[a]	25	25	27	34	33	24

[a] Europe, USSR, North America, Japan, Australia and New Zealand.

Sources: United Nations, *The Determinants and Consequences of World Population Growth*, vol. 1, New York, 1973; J. D. Durand, 'Historical estimates of world population: an evaluation', *Population and Development Review*, 3, 1977, pp. 253–96; United Nations, *Demographic Yearbook 1980*, New York, 1982

Table 4.5 Urban populations, 1950 and 1985

	1950		1985	
	Millions	Percentage of total	Millions	Percentage of total
Developing				
Africa	35.3	15.7	164.6	29.7
Latin America	67.6	41.0	279.4	69.0
Asia	225.8	16.4	791.1	28.1
Total, developing	328.7	17.0	1235.1	31.2
Developed				
North America	106.2	63.9	195.3	74.1
Europe	220.8	56.3	352.2	71.6
Oceania	7.7	61.3	17.5	71.1
USSR	70.7	39.3	182.9	65.6
Total, developed	405.4	53.8	747.9	71.5
World	734.1	29.2	1983.0	41.0

Source: United Nations, *The Prospects of World Urbanization: Revised as of 1984–85*, Population Studies No. 101, New York, 1987

between 1800 and 1900, whereas Egypt's quadrupled, as did Central America's between 1825 and 1900.[11] Thus in 1990 the developing countries contained much the same proportion of the world's population as they did in 1750 (table 4.4). The present high densities and rapid growth are not a sudden event, but the acceleration of a long-term trend.

URBANIZATION AND POPULATION GROWTH

In 1950 the bulk of the population of the developing countries lived in rural areas and were dependent upon agriculture for their livelihood. Since then there has been a radical change in distribution, for cities have grown rapidly both in absolute numbers and as a proportion of the population, the numbers living in towns increasing nearly fourfold and the proportion rising from 17 to 31 per cent (table 4.5). This rapid growth in the urban population has been due both to natural increase in towns, where medical facilities and preventive health are generally superior to those in rural areas, and to migration from the countryside.

The rapid growth of rural population in many parts of the developing world has reduced the size of farms, increased the number without land and accelerated out-migration, for the towns have generally offered higher wages and better social facilities. Unfortunately the growth of employment opportunities in the towns has not kept up with the increase in urban population, so that poverty is widespread in the cities.

This changing distribution of the population has, however, had some significance for the food supply. In the first place, urban incomes in all parts of the developing world are higher than those in the rural areas. Not only do the towns contain most of the manufacturing industry (with its higher wage rates) and a substantial white collar population, but also the government of many developing countries, aware that political power is concentrated in the cities, have attempted to placate the urban population by limiting the increases in food prices. Higher incomes in the towns means that these are the sources of increased demand for animal foodstuffs, and the changes in diet have caused imports of wheat and livestock products even in the poorer developing countries.

Second, the concentration of population in the towns has increased the proportion of food that has to be moved and, because of the high cost of transport in many developing countries, has increased the difficulties of supplying food at reasonable prices.

CHANGES IN THE RURAL POPULATION

Perhaps the most significant contribution of urbanization to agriculture has been to reduce, by migration, the growth of the rural and agricultural populations. After 1800 the process of industrialization was accompanied by rural–urban migration, but in spite of this for much of the nineteenth century the agricultural populations of Europe and the United States continued to increase, for natural increase exceeded out-migration. At the beginning of this century they began a slow decline as out-migration exceeded natural increase. This process has accelerated since the end of the Second World War and agricultural populations in the developed world have more than halved (table 4.6; see also pp. 118–20). In the developing countries, in contrast, the agricultural population has increased by 73 per cent, much lower, of course, than the natural increase of the rural areas. The agricultural labour force has doubled in Asia and Africa, but increased by only a third in Latin America. Some of this increased labour force has been absorbed in the expansion of the cultivated area (see pp. 103–11) or by adopting more labour intensive methods. But in many areas there has been difficulty

Table 4.6 Changes in the agricultural populations of the developing regions, 1950–90

	1950		1990		1950–90	
	Agricultural population[a] (millions)	Agricultural labour force[b] (millions)	Agricultural population[a] (millions)	Agricultural labour force[b] (millions)	Agricultural population[a] (percentage change)	Agricultural labour force[b] (percentage change)
All developed countries	255	147	102	50	–60	–66
All developing countries	1324	529	2287	1040	73	97
Africa	173	66	389	154	124	133
Latin America	87	30	118	41	36	37
Asia	1064	433	1789	860	68	99

[a] Those economically active in agriculture and their dependants.
[b] Those economically active in agriculture.

Sources: FAO, *The State of Food and Agriculture 1973*, Rome, 1973, p. 88; *Production Yearbook 1971*, vol. 25, 1972; *Production Yearbook 1981*, vol. 35, 1982; *Production Yearbook 1990*, vol. 44, 1991

in absorbing these extra numbers, and in many parts of the developing world the subdivision of farms and a rise in the number of landless people has resulted. Without the migration to the towns these problems would have been even more acute.

POVERTY AND HUNGER

In the 1950s there was a fear that the predicted rapid growth of population would exceed the food supply and increase the extent of under-nutrition. There was also a widespread belief amongst economists that increases in the standard of living in the developing countries could only come as a result of industrialization; most thought that rapid population growth would impede this by preventing the accumulation of capital, increasing the number of dependents relative to the workforce and so increasing the expenditure on health and education rather than on productive investment. But this has not been so.

The growth of gross domestic product per capita

Gross domestic product (GDP) is the value of goods and services produced in a country in a year, and the increase per capita in real terms is economic growth. There are many problems in measuring GDP and in making international comparisons of growth. There is no doubt that the GDP of most developing countries was greatly underestimated in the 1950s and the national accounts of socialist countries are prepared in a manner which makes it difficult to compare them with those with market economies.[12] None the less there is little doubt that until the 1980s there was widespread economic growth, even in the poorest areas of the developing world, defined by the World Bank as low income countries (table 4.7). The greatest increases in GDP per capita between 1950 and 1975 came in the Middle East due to oil exports rather than industrialization (table 4.8). However, economic growth was also rapid in the industrializing countries of East Asia such as Taiwan, North and South Korea, Malaysia and of course Japan; it was least rapid in South Asia and Africa.

However, in the 1970s economic growth began to falter in all the developing regions (table 4.9), partly due to the oil crisis of 1972–4. But greater problems were ahead. In the late 1970s many countries in Africa and Latin America borrowed heavily. By the early 1980s they were finding it impossible to pay interest and debts accumulated, reaching a total, for all developing countries, of $600 billion in 1988. Asia was

Table 4.7 Changes in gross domestic product: world (excluding centrally planned economies), 1950–80

	GDP per caput (1980 US dollars)			Average annual percentage growth rate	
	1950	1960	1980	1950–60	1960–80
Industrialized countries	3841	5197	9684	3.1	3.2
Middle income countries	625	802	1521	2.5	3.3
Low income countries	164	174	245	0.6	1.7

Source: World Bank, World Development Report, 1980, Washington, D.C., 1980, p. 35

Table 4.8 Changes in gross domestic product: regions, 1950–75

	GDP per caput (1974 US dollars)		Average annual percentage growth rate, 1950–75
	1950	1975	
South Asia	85	132	1.7
Africa	170	308	2.4
Latin America	495	944	2.6
East Asia	130	341	3.9
China	113	320	4.2
Middle East	460	1660	5.2
All developing countries	160	375	3.4

Source: D. Morawetz, Twenty Five Years of Economic Development 1950 to 1975, Baltimore, Md., 1977, p. 13

least effected by these difficulties and experienced rapid economic growth in the 1980s, but in Africa GDP per capita fell in the 1980s, as it did in Latin America for some of the decade (table 4.9), described by many as the 'lost decade' for economic growth.[13]

None the less the GDP per capita in the 1980s for all developing countries is still well above what it was in the 1950s; the gap between the developed and the developing countries remains (figure 4.1). What has been notable has been the emergence of marked differences between developing countries. Africa and South Asia remain the poorest

Table 4.9 Growth of real per capita gross domestic product, 1965–89

	1965–73	1973–80	1980–9
Middle East and North Africa	5.5	2.1	0.8
East Asia	5.1	4.7	6.7
Latin America	3.7	2.6	0.6
Sub-Saharan Africa	3.2	0.1	−2.2
South Asia	1.2	1.7	3.2

Source: World Bank, *World Development Report 1990*, Washington, D.C., 1990, p. 10

regions, but the Middle East, parts of Latin America and East and South-east Asia have experienced growth and development, and now have substantial manufacturing sectors.

Absolute poverty

But the growth of national real income does not necessarily mean that all sections of the population necessarily improve their standard of living. Indeed some have argued that income inequalities increase in the early stage of economic development, the stage in which most developing countries remain. Unfortunately there is no conclusive evidence on this subject. There is no disagreement on whether income inequalities exist, in either developed or developing countries. Thus in 1970 it was estimated that the poorest 40 per cent of the population of centrally planned economies received 25 per cent of total income, in industrial market economies this proportion was 16 per cent and in developing countries was only 12.5 per cent. Some have argued that inequalities increased in the 1970s, particularly in South Asia, but there has been much disagreement on this, for both the data and the methodology of measurement have been questioned.[14]

Perhaps of more relevance has been the distribution and changing numbers of those suffering from absolute poverty. There have been many attempts to measure absolute poverty. Most have focused upon the minimum income needed to provide a minimum diet; but absolute poverty is a function of low personal incomes, and also of limited access to state provided facilities such as health and education. However, the most recent measure of absolute poverty simply estimates the numbers receiving less than two levels of income per capita (table 4.10).

If $370 is taken as the critical limit, then one-third of the population of the developing countries lived in absolute poverty in 1985. The

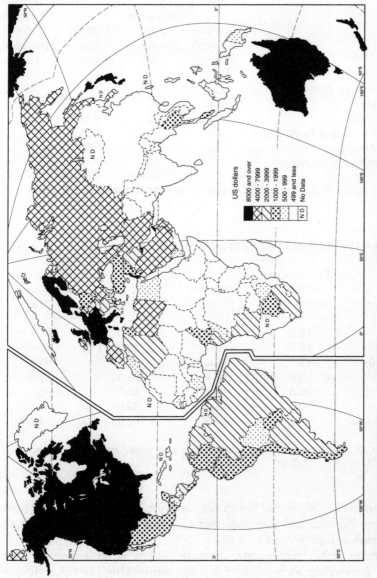

Figure 4.1 Gross national product per caput in US dollars, 1988 *Source*: World Bank, *World Development Report, 1990*, Washington, D.C., 1990

Table 4.10   Number of people with incomes below $275 and $370, 1985

	Incomes below $275		Incomes below $370	
	Number of people (millions)	Proportion of total population (%)	Number of people (millions)	Proportion of total population (%)
Sub-Saharan Africa	120	30	180	47
East Asia	120	9	280	20
South Asia	300	29	520	51
Middle East and North Africa	40	21	60	31
Latin America	50	12	70	19
All developing countries	633	18	1116	33

Source: World Bank, World Development Report 1990, Washington, D.C., 1990, p. 29

Table 4.11   Land and poverty in India, 1970s

Farm size (hectares)	Percentage of rural population	Percentage of land	Per capita income (Rs)	Percentage below poverty line
0	12.3	0	35.0	81.7
0–0.5	18.6	3.0	43.3	75.4
0.5–1.0	15.7	4.0	52.6	67.0
1–2	18.5	12.4	66.8	57.0
2–4	16.3	19.8	85.9	45.3
4–8	10.7	20.4	99.9	31.7
8 and over	7.9	40.0	166.9	4.5

Source: D. Bhattacharya, 'Growth and distribution in India', Journal of Contemporary Asia, **19**, 1989, pp. 150–66

proportion was highest in South Asia and sub-Saharan Africa and least in Latin America and East Asia; but in absolute terms South Asia had the greatest number (table 4.10).

Perhaps rather surprisingly, absolute poverty is far greater in rural areas than urban areas; indeed it was estimated in 1980 that 90 per cent of those living in absolute poverty lived in rural areas. Most of the rural poor are landless or have very small farms.[15]

In India the average income of the rural poor is positively related to

Table 4.12 Per capita food consumption and nutrient intake per day in relation to size of holding, c.1980[a]

Landholding (acres)	Food consumption (g)	Nutrient intake	
		Calories (Kcal)	Protein (g)
0	694	1925	53.9
0.1–0.49	683	1924	52.6
0.5–0.99	745	2035	57.7
1.0–2.99	785	2193	62.5
3.0 and over	843	2375	67.6

[a] Surveys from India, Peru, Bangladesh, the Philippines, Somalia, Haiti, Kenya, Tunisia.
Source: FAO, The Dynamics of Rural Poverty, Rome, 1986, p. 23

Table 4.13 Number of people (millions) with an income of less than US$200

	1950	1960	1970	1972	1977
Gross domestic product data	1178	1249	1226	1177	1043
Consumption expenditure data	1297	1478	–	–	1666

Source: A Berry, F. Bourguignon and C. Morrison, 'Changes in the world distribution of income between 1950 and 1977', Economic Journal, **93**, 1983, pp. 331–50

the size of farm (table 4.11) and a greater proportion of those who are landless or have very small holdings falls below the poverty line. But this relationship between land and income is not confined to India. As food intake is closely related to income, and, in rural societies, income to the size of holding, it is not surprising to find that average calorie consumption in a number of developed countries is closely related to the possession of land (table 4.12).

Earlier it was show that the proportion of the world's population suffering from undernutrition has declined since 1950, although possibly this fall has been arrested in the 1980s. Evidence on changes in the numbers suffering from absolute poverty, or some similar measure of low incomes, is not easily available. An attempt to measure the number with an income of less than $200 produces conflicting results, depending upon the type of data (table 4.13). Using estimates based upon the GDP, the number of poor in 1977 was lower than it was in 1950;

using consumption expenditure data, the number with incomes of less than $200 was 28 per cent higher in 1977 than 1950. Both data show a decline in the proportion of the world's population in this class.

Although the 1980s have seen a decline in the condition of the poor in many parts of sub-Saharan Africa and Latin America, none the less the World Bank has concluded that the condition of the poor was better in the mid 1980s than it was in the 1960s and 1970s; not only are consumption levels now higher but literacy levels are higher and mortality levels lower. There is evidence in Asia at any rate, that the higher national incomes per capita, the better off are the poorest sections of society. It would seem that the proportion of those suffering from absolute poverty has declined in the post-war period, although the absolute numbers may have increased.[16]

CONCLUSIONS

The predictions made in the 1940s and 1950s about the future of world population growth have come true. World population has continued to increase at a very high rate, particularly in the developing countries. Some decline in the rate of increase began in the 1960s in Asia and Latin America, but not in Africa. The gloomy predictions on the effect of rapid population increase upon economic growth, however, have, not been fulfilled; real GDP per capita is universally above the levels of the 1950s in spite of the set-backs of the 1980s. This growth, however, has not eliminated poverty; a substantial proportion of the developing world's population, particularly in sub-Saharan Africa and South Asia, live in absolute poverty. The actual numbers in this condition has probably risen since 1950, but at a far lower rate than the increase of population. That poverty is the primary cause of the persistence of hunger is now rarely questioned. However, the relationships between the growth of food output and the increase in population need to be explored.

5
The Growth of World Food Output

In the immediate aftermath of the Second World War many writers believed not only that there would be very rapid population growth in the forthcoming decades but also that there was little prospect of any substantial increase in agricultural output. It was argued that most of the world's good land was already in cultivation, and any further colonization of new land would only be at prohibitive cost; that the crop yields in Western Europe were already high and could not be greatly increased; and that farmers in the developing countries were primarily subsistence farmers, unresponsive to market forces and unlikely to increase their very low crop yields.[1] Much the same gloomy prophecies had been made in earlier times and indeed are still being made.[2] Before considering how food production has increased since 1950 it may be useful to look briefly at the rate of growth of food production before then.

THE TREND IN FOOD PRODUCTION BEFORE 1950

Little is known of the trends in agricultural production before the eighteenth century; the evidence from tithe returns in Western Europe suggests that food production followed the course of population growth, with output per head rising for brief periods and falling for other periods.[3] In the eighteenth century food output began a steady and continuous increase, and this period has been described by historians as one of agricultural revolution. Even so estimates of the rate of increase suggest that growth was slow; in England and France agricultural output rose at about 0.5 per cent per annum, and so did little more than keep pace with population growth. Indeed, at the time his

Essay was written there was some justification for T. R. Malthus's belief that agricultural output was incapable of keeping up with population growth.[4]

In the nineteenth century output grew more rapidly. England and France had rates of growth over the whole century which exceeded 1 per cent per annum, but output rose more rapidly in the first half of the century than in the second half; this contrast was particularly marked in Britain, where output rose very little between the 1860s and the 1930s. In Germany output rose at 1.3 per cent per annum between 1800 and 1883. In parts of Eastern Europe the end of feudalism and the extinction of open field and common land did not occur until the mid nineteenth century and was followed by rapid agricultural growth – between 1 and 2 per cent per annum in Silesia and Hungary. In the European settled areas overseas the rates of increase were greater than in Western Europe, for much fertile land was brought rapidly into cultivation. In the United States the rate of increase fell below 2 per cent per annum in only two decades, and exceeded 4 per cent in the 1870s. The first half of the twentieth century saw varying rates of growth in agricultural output; in Britain in the 1920s and 1930s output fell, and the depression of the inter-war period retarded the growth of output in other countries.[5]

There are few estimates of the growth of food output in Africa, Asia and Latin America before 1950, for there were few agricultural censuses. There was undoubtedly a rapid development of agricultural output in Latin America, Africa and parts of South-east Asia between the 1870s and the 1930s, but most of this was of crops for export to Europe and the United States. Paul Bairoch has estimated that the output of export crops grew at 2.6 per cent per annum in the late nineteenth century and at over 3.5 per cent between 1900 and 1940. The few estimates of food output suggest much slower growth – notably in India and China – and in the developing countries as a whole food output probably grew more slowly than population.[6]

MEASURING FOOD OUTPUT SINCE 1950

There are two ways of estimating the growth in the volume of food output since 1950. First, estimates of the area in arable land and in the major food crops have been collected and published by the Food and Agriculture Organization (FAO) of the United Nations both for the major regions and for individual countries. Combined with estimates of crop yields these can give some indication of changes in crop production.

Second, indices of agricultural and food output have been prepared and published by the FAO since the early 1950s.

Changes in land use and crop yields

Between 1950 and 1989 the world's arable area increased by 20 per cent, and at a rather greater rate in the developing than the developed region (table 5.1). Within the developed regions, however, the area in arable declined in Europe and showed only a small increase in North America; most of the increase was attributable to the expansion of cultivation in the USSR in the 1950s and to a continuous but smaller growth in Australia. In the developing countries there were substantial absolute and proportional increases in Latin America and Asia, but an apparent decline in Africa. But this demonstrates the limitations of the statistics. Arable land includes not only annual crops such as wheat and permanent crops such as tea or cocoa, but also grass sown for five years or less and fallow. Fallow land is defined as land used for crops but not necessarily in crops every year. It can thus include a variety of different types of fallow. In many semi-arid regions cereals are only sown in alternate years to conserve moisture. In cool temperate areas a cold, late autumn may prevent sowing, and land is recorded as bare fallow. In some countries, notably the United States, land may be withdrawn from crops to prevent over-production, but this idle land is still recorded as arable; nearly 13 million hectares were idle in the United States in 1983, and 8 million in the late 1980s.[7] Most significant, in many parts of the tropics land is sown to crops for two or three years, then abandoned and colonized by the natural vegetation. This period of natural fallow may last for only two or three years or for as long as twenty-five years. In the 1950s and 1960s it was the practice to include such fallow land in statistics for arable, but in the last fifteen years many African countries have ceased to include it in their returns – hence the apparent decline in the area in arable in that continent (table 5.1). A more accurate measure both of the increase in the area in crops and of food output may be obtained by considering the increase first in the area and yield of cereals and second in the area in the major food crops. Both sets of data have their limitations; e.g. some countries record the area *sown*, others the area *harvested*. More seriously in some countries land that is double cropped is returned twice, in others only once. However, the data should show the order of magnitude of increase. Cereal crops not only account for a substantial proportion of the area used for crops, but provide about half the world calorie consumption. Between 1950 and 1988–9 world cereal output more than tripled (table 5.2); that in the developed countries doubled but production in

Table 5.1 Arable land, 1950–88 (million hectares)

	1950	1955	1961–5	1969–71	1975	1989	Change, 1950–89 million hectares	per cent
Developed								
Europe	148	151	152	145	143	140	–8	–5
USSR	175	220	229	232	232	231	57	33
North America	220	229	222	232	253	236	16	7
Oceania	17	25	35	42	47	49	32	189
Total, developed	560	625	638	651	675	656	96	17
Developing								
Latin America	86	102	116	146	154	180	94	109
Asia	348	426	447	443	452	452	104	30
Africa	228	232	190	171	143	186	–42	–18
Total, developing	662	760	753	760	749	818	155	23
World	1222	1385	1391	1411	1424	1474	252	20

Sources: FAO, *Production Yearbook 1981*, vol. 35, Rome, 1982, pp. 45–56; *Production Yearbook 1976*, vol. 30, Rome, 1977, pp. 45–56; *Production Yearbook 1957*, vol. 11, Rome, 1958, pp. 3–7; *Production yearbook 1990*, vol. 44, 1991

Table 5.2 Estimates of the increase in the volume of output from 1948–52 to 1988–9: cereal crops only

| | 1948–52 | | | 1988–9 | | | Increase from 1948–52 to 1988–9 | | |
	Area (million hectares)	Yield (tonnes per hectare)	Production (million tonnes)	Area (million hectares)	Yield (tonnes per hectare)	Production (million tonnes)	Area (%)	Yield (%)	Production (%)
World	511	1.13	578	702	2.5	1782	37	121	208
Developed	293	1.29	378	280	2.9	812	–4	125	115
Developing	218	0.92	200	422	2.3	970	94	150	385

Sources: T. N. Barr, 'The world food situation and global grain prospects' *Science*, **214**, 1981, pp. 1087–95; FAO, *Production Yearbook 1989*, vol. 43, Rome, 1990

the developing countries rose fivefold. This increase has been achieved by increases both in the area sown to cereals and in yields.

Cereals, however, are not the only source of calories; consequently increases in the area in the major food crops – cereals, roots and tubers, pulses and oilseeds – give an alternative measure of the increases in the food supply. This area increased by 26 per cent between 1950 and 1988 (table 5.3). If it is assumed that the yield of all food crops increased at the same rate as the cereals then food output rose by 160 per cent between 1950 and 1988 (table 5.4), a lower rate than the increase in cereals but well in advance of the increase in population over the same period. But apart from the obvious limitations of these estimates neither includes food derived from livestock, a major source of calories in the developed world, and of increasingly importance in the richer developing countries. Fortunately the indices of food production compiled by the FAO do include both crop and livestock production.

Indices of food output

The second way of measuring the growth of world and regional food output is to use the indices of food production constructed by the FAO; they have been published for all countries annually since 1960, and for the major regions since 1948–52, with the exception of China, which is only included after 1960.

There are considerable problems in preparing these indices. First of all, many countries do not take agricultural censuses at regular intervals; indeed many still have to hold an agricultural census. Second, most agricultural censuses record the area in crops and sometimes the yield, but not the volume of output. Similarly agricultural censuses may record the number of livestock in a country, but not the output of milk or meat. The latter will only be available if the records of slaughterhouses are known. Third, in calculating output for a given year it is important to estimate net output and to avoid double counting. Thus, for example, a part of any crop harvest has to be retained for seed for the following year. In developed countries much of the cereal output is fed to livestock; it must not be counted twice. Nor are all food crops used for human consumption; oilseeds, for example, may be used for industrial purposes. A fourth problem is that to quantify total food output quite different commodities – such as eggs and wine, olive oil and pineapples – have to be converted into comparable units of account.

Ideally all foods should be reduced to their calorific value, so that the growth of food output measures the amount of food available and allows an easy comparison with nutritional requirements. But whereas the FAO has published comprehensive data on the number of calories

Table 5.3 Area in major food crops,[a] 1948–88 (million hectares)

	1948–52	1961–5	1969–71	1979–80	1988	Change, 1948–88	
						million hectares	per cent
Developed							
North America	124.3	98.1	101.2	126.9	116.6	−7.7	−6.2
Europe	92.2	92.5	88.7	86.3	84.7	−7.5	−8.1
USSR	108.5	145.0	137.3	141.8	128.0	19.5	17.9
Oceania	6.6	9.9	12.7	17.2	16.6	10.0	151.5
Total, developed	331.6	345.5	339.9	373.2	345.9	14.3	4.3
Developing							
Latin America	43.5	57.5	69.5	86.7	91.3	47.8	109.8
Asia	337.7	385.0	375.0	393.0	420.9	83.2	24.6
Africa	57.8	80.2	89.9	98.9	114.2	56.4	97.5
Total, developing	439.0	522.7	534.4	578.6	626.4	187.4	42.7
World	770.6	868.2	874.3	950.8	972.3	201.7	26.2

[a] Includes all grain crops, potatoes, sweet potatoes, yams, pulses and oilseeds.
Sources: FAO, *Production Yearbook 1981*, vol. 35, Rome, 1982, pp. 93–137; *Production Yearbook 1976*, vol. 30, Rome, 1977, pp. 89–134; *Production Yearbook 1957*, vol. 11, Rome, 1958, pp. 31–2; *Production Yearbook 1988*, vol. 42, Rome, 1988

Table 5.4 Estimates of the increase in the volume of output from 1948–52 to 1988: total area in major food crops multiplied by average yield of cereals[a]

	1948–52			1988			Increase from 1948–52 to 1988		
	Area (million hectares)	Yield (tonnes per hectare)	Production (million tonnes)	Area (million hectares)	Yield (tonnes per hectare)	Production (million tonnes)	Area (%)	Yield (%)	Production (%)
World	771	1.21	933	972	2.5	2430	26	107	160
Developed	332	1.53	507	346	2.9	1003	4	90	98
Developing	439	0.97	425	626	2.3	1427	42	137	236

[a] Food crops other than cereals are assumed to have increased their yields at the same rate as cereals.

Sources: FAO, Production Yearbook 1957, vol. 11, Rome, 1958; Production Yearbook 1967, vol. 21, Rome, 1968; Production Yearbook 1976, vol. 30, Rome, 1977; Production Yearbook 1981, vol. 35, Rome, 1982; Production Yearbook 1989, vol. 43, Rome, 1990

available in each country from output, stock and imports (see pp. 17–19), the indices of food production are not calculated on such a basis. Instead output is based on physical output weighted by local prices and, to ensure international comparability, all values of national outputs are expressed in US dollars. FAO statisticians have prepared long-term indices for food production. However, these are not published in terms of value; instead, output for any given year is related to output in a base period which is expressed as 100. These indices are the source of most statements on the rate of increase of food production over the last forty years.[8]

RATES OF INCREASE IN WORLD FOOD OUTPUT, 1950–86

World food output is estimated to have increased at 3.1 per cent per annum in the 1950s, 2.7 per cent in the 1960s and 2.5 per cent in both the 1970s and the 1980s (table 5.5(a)). These figures are well above the rates of increase recorded in individual countries in Europe in the eighteenth and nineteenth centuries, or indeed in the first half of this century; on the other hand they are lower than those recorded in the United States or other land abundant countries in the nineteenth century. It should be noted, however, that the rate of increase has declined from the high point in the 1950s.

There have been important regional differences, however, in the rate of growth. First, it has been consistently higher in the developing regions than in the developed. This became particularly noticeable in the 1980s when increase in the developing regions was higher than in the past, whilst in the industrial countries measures to reduce surplus production in both the European Community and the United States have had an effect. Second, Africa has had a lower rate of increase than any other developing region throughout the period. Third, the Near East and Latin America have experienced a fall in the rate of increase in the early 1980s.

It is clear from the data presented in this chapter so far that food output has increased rapidly in the post-war period; not only have rates of increase exceeded anything achieved in the past, but developing regions have increased more rapidly than developed. Between the early 1950s and the late 1980s world cereal output, which accounts for half the world's calorific intake, has tripled; the output of the major food crops has been estimated to increase by 160 per cent; and the FAO *agricultural* index in the early 1980s was 159 per cent higher than in the 1930s, and 135 per cent higher than in 1950.[9]

Table 5.5   Food output, 1952–4 to 1986

	1952–4 to 1959–61	1961–70	1971–80	1980–6
(a) Rate of increase in food output from 1952–4 to 1986 (per cent per annum)				
Africa	2.1	2.7	1.8	2.1
Far East	3.4	3.5	3.6	3.5
Latin America	3.1	3.5	3.8	2.2
Near East	3.3	3.0	3.5	2.1
Asian CPEs	No data	2.7	3.2	5.4
All developing countries	3.1	3.1	3.3	3.5
All developed countries	3.0	2.4	1.9	1.5
World	3.1	2.7	2.5	2.5
(b) Rate of change in food output per caput, 1952–4 to 1980 (per cent per annum)				
Africa	−0.2	0.1	−1.2	
Far East	1.1	0.9	0.9	
Latin America	0.3	0.8	1.2	
Near East	0.8	0.3	0.6	
Asian CPEs	No data	0.9	1.6	
All developing countries	0.7	0.7	1.0	
All developed countries	1.7	1.4	1.1	
World	1.1	0.8	0.6	

[a] CPEs, centrally planned economies.
Sources: FAO, *World Agriculture: The Last Quarter Century*, Rome, 1970, p. 9; *The State of Food and Agriculture 1981*, Rome, 1982, pp. 5–6; *The Fourth World Food Survey*, Rome, 1977, p. 4; R. Maurio, 'Balanced strategy for food producing self-reliance', *Cooperazione*, 85, 1989, supplement, p. 30

CHANGES IN FOOD PRODUCTION PER CAPUT

World food output has more than doubled since 1950. But of course this praiseworthy achievement by farmers must be set against the very rapid growth in population since 1950. Between 1950 and 1990 the population of the world doubled; but this was significantly *less* than the increase in world food output. J. L. Simon has calculated that world food output per capita in 1976 was 28 per cent above that in 1950 and by the 1980s it was probably more than a third above that in 1950.[10] There is thus no question, at the global scale, of population having outrun food production since 1950.

But, of course, there has been a fundamental regional difference in the rate of population growth. In the developed world in the period 1950–90 the population increased by 36 per cent. In contrast, in the same period the population of the developing regions rose by 138 per cent; Latin America increased its population by 172 per cent, Africa by 186 per cent and Asia by 126 per cent (table 4.1). The consequences of this differential population growth are clear. Until 1980 food output per caput rose markedly in the developed world, where considerable food increases were combined with modest population growth; but in the developing world dramatic increases in food production were matched by equally dramatic increases in population (figure 5.1). None the less, between 1950 and 1980 food production kept ahead of population growth in the major regions, and by 1980 in the Near East, the Far East and Latin America food output per caput was some 10–15 per cent above the level of the 1950s (figure 5.2). This was not so in Africa, however, where the rate of population increase in the 1970s was *above* that of the 1950s and 1960s, but the rate of increase in food production fell considerably. Hence output per caput fell from the early 1960s to 1980 (table 5.5 and figure 5.2). In the 1980s the developing regions had mixed fortunes. Output per capita continued its decline in Africa (figure 5.3), the upward trend in the Near East was reversed and Latin America held its own. But the upward trend in the Far East continued, and in China was dramatic.

THE STRUCTURE OF FOOD OUTPUT

In the last forty years nearly all of Europe's extra food output has come from an increased yield per hectare; the area in arable has declined in every country in Europe since 1950. In contrast much of Latin America's increased output has come from extending the area in cultivation, and yield increases have been less important. Historically the relative importance of yield and area has varied considerably.

There is little reliable information upon trends in land use, yields or output before the nineteenth century, but most of the increase in output probably came from bringing new land into cultivation, for yields were low and showed little increase over time. Thus in most of Europe crop yields increased by no more than 10 per cent between 1500 and 1800. The exception was in England and the Low Countries, where yields increased by one-third or more, mostly after 1600.[11]

Most historians believe that the later eighteenth and nineteenth centuries saw marked increases in crop yields, and this period has been described as one of agricultural revolution. Higher yields were obtained

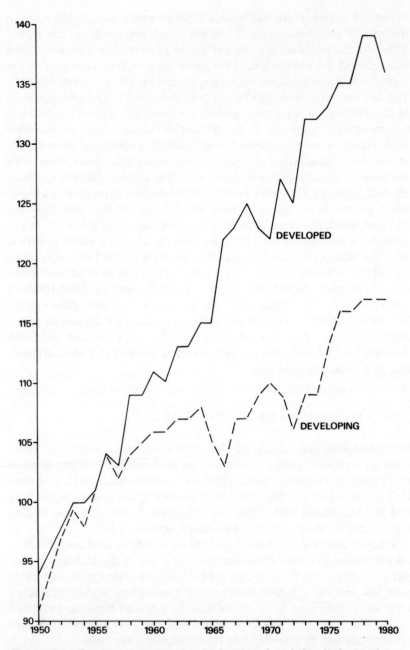

Figure 5.1 Food output per caput in developed and developing regions, 1950–80; 1952–6 = 100; China is excluded before 1960 *Sources*: FAO, *Production Yearbook 1981*, vol. 35, Rome, 1982; *Production Yearbook 1976*, vol. 30, Rome, 1977; *Production Yearbook 1970*, vol. 24, Rome, 1971

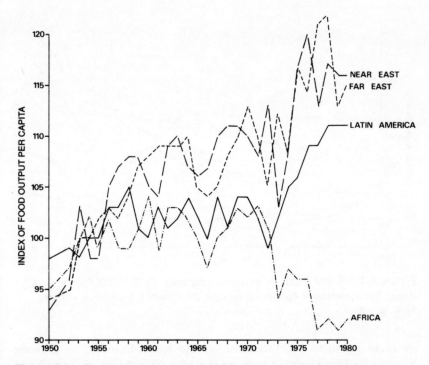

Figure 5.2 Food output per caput, 1950–80, in developing regions (excluding China); 1952–6 = 100 *Source*: FAO, *Production Yearbook 1970*, vol. 24, Rome, 1971; *Production Yearbook 1981*, vol. 35, Rome, 1982

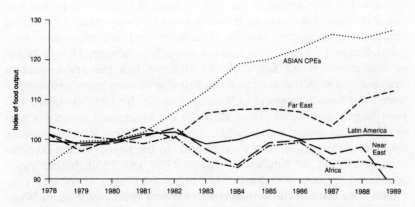

Figure 5.3 Food output per caput in developing regions, 1978–89; 1979–81 = 100 *Source*: FAO, *Production Yearbook 1989*, vol. 43, Rome, 1990

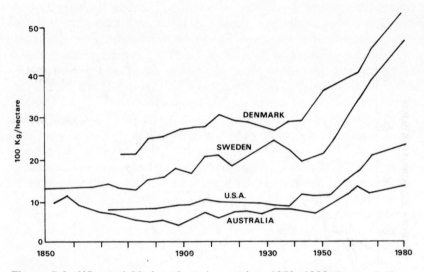

Figure 5.4   Wheat yields in selected countries, 1850–1980 *Source*: D. B. Grigg, *The Dynamics of Agricultural Change: the Historical Experience*, London, 1982, p. 130

by more intensive cultivation to rid the land of weeds, by growing legumes such as clover to increase the nitrogen content of the soil and by feeding livestock on roots, hay and oilseeds to increase both their weight and the manure supply.[12] Higher yields were also obtained by growing new crops with a higher calorific output per hectare such as potatoes, sugarbeet and maize. Thus output per hectare in Western Europe rose steadily if not dramatically in the nineteenth century, and continued to increase up to the 1930s (figure 5.4). But this period also saw a continued increase in the area sown to crops. After 1750 rising cereal prices prompted farmers to reclaim land not previously cultivated. But equally important farmers throughout Western Europe began to sow crops on the fallow. In 1700 about half the arable land in England was in fallow each year, and similar or even larger proportions were to be found elsewhere in Western Europe. By 1900 the fallow had been largely eliminated. Thus although the nineteenth century is normally seen as one of rising yields, the increased area in crops made a major contribution to extra food output. In the heyday of the 'agricultural revolution' in England, between 1750 and 1850, higher yields accounted for only one-third of the increased output, extra area two-thirds.[13] By the end of the nineteenth century most countries in Western Europe had reached their maximum arable area, and there was little change until 1950. Since the 1950s there has been a decline in arable area, but grain yields have doubled. Thus, in Western Europe there has been a progressive increase in the contribution of crop yields to the raising of food production. Before the eighteenth century increased yields made very little contribution, and until the mid nineteenth century colonizing new land and reducing the fallow accounted for the majority of output increases. By the 1980s yield increases were accounting for all but a small proportion of output increases.

The European settlements overseas – Canada, Australia, the United States – together with the Soviet Union have had a very different history, owing largely to their late settlement and lower population densities. In 1800 Western Europe had a long history of settlement and high population densities, as did parts of northern Russia; but southern Russia and the west of North America and Australasia were sparsely populated although containing large areas of fertile land. These were settled in the nineteenth century, and it was this addition of new land that accounted for most of their increase in food output. Crop yields showed very little sign of increase in the nineteenth century or indeed in the first thirty or forty years of this century (figure 5.4), and wheat yields remained well below those in Western Europe. Since the 1930s, however, crop yields in North America have increased dramatically, and have accounted for most of the increase in food output since 1950. Only in the Soviet Union and Australia have increases in the area sown to crops accounted for much of the increase in output.

YIELD AND AREA EXPANSION IN FOOD OUTPUT INCREASES SINCE 1950

Since 1950 yield increases have accounted for a growing proportion of total food output. The evidence for this is provided by data upon cereals alone (table 5.6), and it should be recalled that other crops locally provide an important part of the diet. In sub-Saharan Africa, for example, roots provide 35 per cent of the calories consumed.[14] Furthermore the increase in the yield of cereals has been greater than increases in the yield of other food crops. None the less they provide an approximate guide to events between 1950 and 1980.

First, it can be seen that increases in yield accounted for more than four-fifths of the world increase in cereal production between 1950 and 1980. In the 1950s the contribution of area expansion was greater, and has since declined. Second, in the developed countries yield increases have been of overwhelming importance, although in the 1970s some of the additional output did come from area increases, mainly in the United States (see pp. 117–18). In the developing countries, on the other hand, some 40 per cent of all increased cereal output in 1950–80 came from

Table 5.6 Relative contribution of yield and area to increases in world cereal output, 1950–80 (percentages)

	1950–60		1960–70		1970–80		1950–80	
	Area	Yield	Area	Yield	Area	Yield	Area	Yield
Developed	−2	102	−5	105	42	58	3	97
Developing	82	18	25	75	16	84	40	60
World	56	44	12	88	25	75	15	85

Source: T. N. Barr, 'The world food situation and global grain prospects' *Science*, **214**, 1981, pp. 1087–95

bringing new land into cultivation; yield increases, however, have grown progressively more important, in the 1980s yield increases accounted for nearly all the extra output in both the developed and developing world.[15]

There were important regional variations in the contributions of area expansion to the growth of cereal output in the 1960s and 1970s (table 5.7). In the developing world increases in the cultivated area were of most importance in Latin America and Africa, both comparatively sparsely populated and said to contain large reserves of cultivable land. In contrast in South Asia and the Middle East most production increases came from yield increases. But these figures probably understate the importance of area expansion. Thus it has been estimated that area expansion accounted for two-thirds of the increase in agricultural output in Latin America between 1950 and 1980; most authorities believe that there has been little increase in crop yields in Africa and that the expansion of the cultivated area has accounted for most of the extra food output over the last forty years.[16]

CONCLUSIONS

It has been shown in this chapter that there have been considerable increases in food output in the world since 1950 – output more than doubled – and that the rate of increase has been higher in the developing than the developed countries. However, the much higher rates of population growth in the developing world have meant that there has been comparatively little increase in output per caput; food production has just keep ahead of population growth. In some countries – mainly in Africa – there has been a decline in output per head over

Table 5.7 Cereal production increases due to area expansion and yield increase from 1961–2 to 1979–80, by major regions (percentages)

	Area	Yield
Developing		
Latin America	54	46
Africa	52	48
East Asia	48	52
South Asia	27	73
Middle East	20	80
Total, developing	23	77
Developed		
United States	19	81
Oceania	13	87
Soviet Union	4	96
Western Europe	0	100
Eastern Europe	−21	121
Total, developed	14	86
World	19	81

Source: T. N. Barr, 'The world food situation and global grain prospects', *Science*, **214**, 1981, pp. 1087–95

the last thirty years. Hence the great regional differences in per caput consumption already apparent in the 1930s and 1948–52 have been perpetuated.

Until the middle of the nineteenth century much of the increased food output required by slow population growth came from colonizing new land or, particularly after 1600, by reducing the fallow area. From the middle of the nineteenth century, however, crop yields began to increase more rapidly and in Western Europe have accounted for nearly all the increased food output since 1950. In the developing countries area expansion has continued to be an important source of extra food production, particularly in Africa and Latin America, but since the mid 1960s much of the extra food in these regions has come from higher yields.

The following five chapters examine in more detail how food output has been increased since 1950. Chapter 6 reviews the expansion of the cultivated area in different parts of the world; chapters 7–10 look at the expansion of food production in the developed world and in Africa, Latin America and Asia.

6

The Expansion of the World's Arable Land

It was not until the middle of the nineteenth century that increases in crop yields began to contribute a significant proportion of the increases in food production. Before then in Europe and Asia the expansion of the arable area was the prime response to the slow rise in the population. The more rapid increases in population in the nineteenth century led to not only higher crop yields but also the need for food imports. Europe's arable area stagnated from the late nineteenth century until the 1940s and then began to decline. Most of the food imports into Europe came from the European settled areas overseas and from southern Russia. Between the middle of the nineteenth century and the Second World War there was a remarkable increase in the area in cropland in these countries, particularly in the United States and Russia. Over 200 million hectares of new cropland came into cultivation between 1860 and the 1930s (table 6.1). Estimates of the area in arable in the developing world before this century, or indeed before the 1950s, are unreliable. Nevertheless those estimates that are available suggest a steady expansion of the area in crops in Japan, China, India and Java; furthermore arable expansion probably kept up with population growth until the late nineteenth century, since when there has been a decline in arable per caput in all but India (table 6.2).

The world's arable area may have increased by two-thirds or more in the period between 1870 and 1950.[1] By the latter date most of the world's good land seemed to be in cultivation, and few writers then believed there were large areas left for farmers to colonize. Yet since 1950 the world arable area has increased by 20 per cent (table 5.1), some 250 million hectares, more than the area added in the overseas European settlements between 1870 and 1930 (table 6.1).

Table 6.1 The expansion of arable in areas of recent settlement

	1860	1880	1900	1920	1930	1960	1989
United States	65.8	75.9	128.8	162.4	166.8	158.3	190.0
Russia	49.2	102.6	113.3	–	109.4	195.9	231.0
Canada	–	6.06	8.4	20.2	23.4	25.0	46.0
Argentina	–	–	5.6	–	24.2	22.0	36.0
Australia	0.4	1.6	3.2	6.0	10.1	11.7	49.0

Source: D. B. Grigg, *The Agricultural Systems of the World: An Evolutionary Approach*, Cambridge, 1974, p. 262; FAO, *Production Yearbook 1990*, vol. 44, Rome, 1991

THE POTENTIAL ARABLE AREA

Much of the new land settled after 1860 was very fertile, particularly in the United States corn belt and the Russian chernozem zone. Once these areas had been brought into cultivation it was thought that there was little good land left. However, it has always proved remarkably difficult to predict the supplies of potential arable land. Unused land can be classified on the basis of its physical characteristics. Some land, such as the permafrost areas of North America or northern Eurasia, is unlikely ever to be used for crops; nor are desert areas without access to irrigation water, or the steep slopes of much of the world's mountains. The cultivation of the rest depends upon factors that are difficult to predict, such as future technology, food prices and production costs.

Not surprisingly, estimates of the potential arable area vary greatly (table 6.3). Thus Pearson and Harper estimated in 1945 that the area suitable for food crops was only 1048 million hectares, which was less than the area then under arable. At the other extreme C. B. Fawcett and L. D. Stamp put the potential arable at over 4000 million hectares, three times the present arable area. Most estimates of the potential arable have been made by calculating the areas too dry, too cold, too steep and with infertile soils and by assuming that the residual could be cultivated. Two recent estimates, however – those of the President's Advisory Committee and of P. Buringh and a group of Dutch agronomists – have built up their world estimates from a consideration of small regions based on soil and climatic characteristics. Their estimates are remarkably similar: they argue that about 3200 million hectares could be used to grow crops, one-quarter of the world's land area; as some 1400 million hectares are in use, 1800 million hectares remain to be cultivated.

Table 6.2 The arable area and arable land per caput of the total population, selected Asian countries

	Japan		China		Indian subcontinent		Java	
	Arable area (hectares)	Arable per caput (hectares)	Arable (million hectares)	Arable per caput (hectares)	Arable (million hectares)	Arable per caput (hectares)	Arable (million hectares)	Arable per caput (hectares)
1600	2.1	0.12	33.5	0.21	46.4	0.37	–	–
1800	3.0	0.12	63.4	0.23	–	–	1.5	0.3
1870	3.5	0.12	81.6	0.23	–	–	–	–
1900	5.4	0.12	91.7	0.21	79.5	0.28	6.6	0.23
1920	5.9	0.11	–	–	–	–	8.0	0.23
1930	6.0	0.09	98.9	0.19	83.2	0.21	8.4	0.2
1960	5.7	0.06	113.1	0.17	151.5	0.26	8.8	0.14
1988	4.7	0.04	97.0	0.1	197.8	0.2	n.d	n.d

Sources: D. B. Grigg, *The Agricultural Systems of the World: An Evolutionary Approach*, Cambridge, 1974, pp. 84, 92, 96, 100, 261; FAO, *Production Yearbook 1989*, vol. 43, Rome 1990

Table 6.3 Estimates of the potential arable area (million hectares)

	Total land (a)	Potential arable (b)	Arable in use (c)	Additional arable (d)	(b) as a percentage of (a)	(d) as a percentage of (c)
Fawcett, 1930	14,510	4404	–	–	29	–
Alsberg, 1937	14,510	2591	1424–554	1037–167	18	67–80
Pearson and Harper, 1945[a]	14,450	1048	619	429	7	69
Baker, 1947	13,470[b]	1554	1036	518	12	50
Salter, 1947	14,500	1994	1015–450[c]	526	14	36–52
Kellog, 1951	–	1615	1012	603	–	59
Stamp, 1952	14,454	4084	1214–619	2465–870	28	152–236
Prasslov and Rasov, 1950	14,500	3700	1384[d]	2316	26	167
Orvedal, 1958	14,812	4027	1384	2643	27	190
President's Committee, 1967	13,150[b]	3187	1386	1801	24	130
Buringh et al., 1974	13,530[b]	3220	1413	1807	24	128
Gerasimov, 1983	13,339	2678	1427	1251	20	88

[a] Food crops only.

[b] Excluding polar regions.

[c] Salter's figure for land in arable is 7–10 per cent of land surface, for which he gives no estimate but appears to follow Prasslov and Rasov.

[d] Prasslov and Rasov's estimate of arable land was 783 million hectares; 1384 million hectares is the FAO estimate for 1950.

Sources: C. B. Fawcett, 'The extent of the cultivable land', *Geographical Journal*, **76**, 1930, pp. 504–9; C. Alsberg, 'The food supply in the migration process', in I. Bowman (ed.), *Limits of Land Settlement*, Washington, D.C., 1937, pp. 25–6; F. A. Pearson and F. A. Harper, *The World's Hunger*, New York, 1945; O. E. Baker, 'The population prospect in relation to the world's agricultural resources', *Journal of Geography* **46**, 1947, pp. 203–20; R. M. Salter, 'World soil and fertiliser resources in relation to food needs', *Science*, **105**, 1947, pp. 533–8; C. E. Kellog, *Food, Soil and People*, London, 1950; L. D. Stamp, *Land for Tomorrow*, Bloomington, Ind. 1952; Prasslov and Rasov quoted in L. D. Stamp, *Our Developing World*, London, 1960, pp. 63–4; A. C. Orvedal quoted in US Department of Agriculture, Yearbook for 1964, *Farmers World*, Washington, D.C., 1965, pp. 62–3; President's Science Advisory Committee, *The World Food Problem*, vol. 2, Washington, D.C., 1967, p. 433; P. Buringh, H. D. J. Van Heemst and G. J. Staring, *Computation of the Absolute Maximum Food Production of the World*, Wageningen, 1975; P. Gerasimov, 'Land resources of the world, their use and reserves', *Geoforum*, **14**, 1983, pp. 427–39.

Table 6.4　Arable land and potential arable land, 1967

	Total area (a)	Potential arable area (b)	Arable land in use (c)	Additional arable land [(b) − (c)] (d)	Potential arable land as percentage of total area (e)	Additional land as percentage of arable land in use [(d) as percentage of (c)] (f)
Developing						
Africa	3,019	732	158	574	24.2	363
Asia	2,735	627	518	109	22.9	21
South America	1,752	679	77	602	38.8	781
Total, developing	7,506	2,038	753	1,285	27.2	170
Developed						
Europe	478	174	153	21	36.4	14
Australia	822	154	16	138	18.7	862
North America	2,110	456	238	227	22.0	95
USSR	2,234	356	226	130	15.9	58
Total, developed	5,644	1,149	633	516	20.4	82
Total	13,150	3,187	1,386	1,801	24.2	130

Source: President's Science Advisory Committee, *The World Food Problem*, vol. 2, Washington, D.C., 1967, p. 434

Of the 1800 million hectares that remain to be brought into cultivation according to the President's Science Advisory Committee (table 6.4), just over one-quarter are in the developed world and 71 per cent in the developing countries. Of the former, Europe has very little potential, North America the most. In the developing regions Asia has little potential, indeed only one-fifth more than the present arable area. Thus nine-tenths of the potential arable in the developing regions – and two-thirds of the world's total – lie in the tropical regions of Africa and South America, in the rain forest and savannah zones of those two continents. Although the world – in particular the tropical regions of

Africa and Latin America – apparently has large reserves of future crop-land, two caveats must be made. First, although only 11 per cent of the land area of the earth is used for crops, some 24 per cent is used for grazing livestock; the latter area corresponds very closely to the areas of potential but unused arable land. The conversion of these areas to crops would of course result in a net gain in food output, but not quite as large a gain as might be initially supposed. Second, although considerable parts of the rain forest and the savannah areas of Latin America and, especially, Africa are already in cultivation, these regions do present technical problems not yet resolved.[2]

THE DISTRIBUTION OF ARABLE LAND

In view of the great concern over world food supplies, it is somewhat surprising to find that only 11.3 per cent of the world's land area is used for arable, and not all that is used for crops every year. Although some arable land is found in nearly all parts of the globe, there are four great zones which contain two-thirds of the world total (table 6.5). The first is in North America, mainly west of the Appalachians, and reaching north into the Canadian prairies (figure 6.1). The extension of this zone is limited northwards by the shortness of the growing season and westwards by low and variable rainfall, although there has been an increase in irrigation in the west in the last forty years. Most of this area has been settled only since 1800; crop yields are low when compared with Western Europe but output per head of the labour force is high.

Second is the great zone stretching from Ireland to the Urals and beyond; it includes one-quarter of the world's arable land compared with 16 per cent in North America, but supports a far greater population. To the north cold winters and short growing seasons preclude cultivation, and the considerable upland areas are used mainly for grass rather than crops. Most of the region has a long history of agricultural settlement although the great frontier area of the chernozem soils of southern Russia has been occupied only since the late eighteenth century and the semi-arid region east of the Urals was not brought into cultivation until the 1950s. There is a marked difference between west and east. In Eastern Europe and the Soviet Union there is 0.7 hectare of arable per head of the total population, not far short of that in North America, but in Western Europe there is only one-quarter of a hectare, little more than that available in the Far East (table 6.5).

In the developing world the two great zones are the Indian subcontinent and China; both areas have very long histories of settlement and both have high population densities, so that there is very little arable land per caput. In Africa and Latin America there are no great continuous

Table 6.5 Arable land, 1988

	Million hectares	As a percentage of land area	As a percentage of world area	Hectares per caput total population
Developed				
North America	236	12.2	16.2	0.87
Western Europe	94	25.4	6.4	0.25
Australasia	47	6.0	3.2	2.4
USSR and				
Eastern Europe	278	11.9	19.0	0.7
Japan	5	12.4	0.3	0.04
Total, developed	660	12.3	45.2	0.53
Developing				
Africa	156	6.7	10.6	0.32
Latin America	180	8.9	12.3	0.42
Near East	83	6.9	5.6	0.31
Far East	270	33.5	18.5	0.18
Asian CPES	110	9.6	7.5	0.09
Other developing	1	1.4	–	0.5
Total, developing	800	10.6	54.8	0.2
World	1460	11.3	100.0	0.29

CPEs, centrally planned economies.
Source: FAO, *Production Yearbook 1989*, vol. 43, Rome, 1990

zones of cropland as found in the four regions earlier described, but they are both apparently well endowed with arable land, containing together one-fifth of the world's arable land and possessing respectively 0.3 and 0.4 hectares per head. But as will be seen, much of this arable land is infrequently cropped (see pp. 98–9).

Although Western Europe and Japan have very low supplies of arable land per caput, the developed world as a whole has 45 per cent of the world's arable land but less than one-quarter of the population. This is a major, obvious, but neglected factor in accounting for the world distribution of food supplies.

THE EXPANSION OF THE ARABLE AREA, 1950–88

Earlier it was noted that the world's arable area had increased by 20 per cent between 1950 and 1989 (table 5.1) and the area in the major food

Figure 6.1 The world distribution of arable land *Source: The Times Atlas of the World*, London, 1968, pp. XXVI–XXVII

crops by 26 per cent (table 5.3), and that this increase had been greatest in Latin America and Africa and least in Asia. These figures almost certainly understate the actual expansion of the area in crops.

There are three ways in which the area in crops can be increased. The first is the *reduction of the fallow*. As noted earlier, many farming systems, notably in Africa and Latin America, have a period of natural fallow in which soil fertility is restored under the natural vegetation. In the last thirty or forty years the length of the fallow period has been shortened and the period in crops has increased, much as it was in Western Europe between 1650 and 1900. Second, in most European farming systems only one crop can be grown in a year; the growing season is generally too short for more than one staple food crop. However, in much of the tropics it is possible to grow two major crops one after the other in one year. The increase of *multiple cropping* has probably been of considerable importance in increasing the area sown to crops in Asia since 1950. Third, land hitherto unused, *new land*, has been colonized in many parts of the world since 1950, although often at the expense of grazing land. Last, it should be noted that these considerable increases in the area sown to crops have not been a complete net gain, for there has been a loss of cropland to urban expansion, soil erosion and salinity.

The reduction of fallow

In the eighteenth and nineteenth centuries the fallow in Europe was drastically reduced. Sown with potatoes and sugar-beet it directly increased food output; sown with clover and fodder roots it increased the weight and number of livestock and the supply of manure. But in the developing world much of the arable area is still in fallow. In the 1980s (table 6.6) it was estimated that 68 per cent of the arable land in North Africa and the Middle East was sown to crops in any one year: rainfed cereals need a fallow between crops to conserve soil moisture, and in much of the irrigated area salinity and waterlogging requires land to be temporarily withdrawn from cultivation.[3] In tropical Africa much of the arable is still farmed by shifting agriculture and bush fallowing: land is sown to crops for two or three years and then abandoned to the natural vegetation for periods from three to thirty years. If the period of regeneration is long enough – it varies according to the climate and type of vegetation – soil fertility is restored and the land may be cropped again. Throughout tropical Africa rising population densities have caused the reduction of the fallow and an increase in the area in crops. However, if no alternative means of maintaining soil fertility – such as

Table 6.6 Cropping intensities, 1960s and 1980 (percentage of arable land sown to crops in a year)

	1960s	1980
Near East and North Africa	56	68
Tropical Africa	42	55
Latin America	54	62
Asia[a]	100	107

[a] Excluding South-west Asia.
Sources: FAO, *Provisional Indicative World Plan for Agricultural Development*, vol. 1, 1970, p. 45; J. P. Hrabovszky, 'Agriculture: the land base', in R. Repetto (ed.), *The Global Possible, Resources, Development and the New Century*, New Haven, Conn., 1985, pp. 211–55

growing leguminous crops or the use of manure or chemical fertilizers – is adopted, soil fertility will decline, and eventually soil erosion may cause the land to be abandoned. There is no doubt that this process of fallow reduction has occurred, has increased the area in crop, and in some cases has led to the degradation and destruction of land, but there are no reliable comprehensive statistics (see pp. 112–13). In the 1960s, however, only 42 per cent of the arable of tropical Africa was thought to be in crops but by 1980 this had reached 55 per cent.[4] In Latin America considerable areas are in natural fallow and much the same process of fallow reduction and later degradation has occurred. There are also areas of grain production in semi-arid zones where fallowing is practised, and the irrigated areas, as in the Middle East, are underutilized. Only half Mexico's considerable irrigated area is cropped in any one year. In 1980 it was estimated that 62 per cent of Latin America's arable was sown to crops. In Asia, in contrast, 107 per cent of the cropland was sown each year. But this was an average reflecting very great regional variations. In the sparsely settled forest areas of Borneo or the interior of South-east Asia shifting cultivation is still to be found, where fallows may last for twenty-five years. In contrast, in many parts of China, the Indian deltas and plains and Java and the Philippines, much of the land carries two grain crops in a year.[5]

Much of the world's arable still lies in fallow each year – one-third in the 1970s according to P. Buringh[6] – but the fallow has been continuously reduced in the post-war period as the population and the need for extra food has grown. By the 1980s the area in fallow in the developing countries was less than in the 1960s (table 6.6).

Table 6.7　Estimates of arable area multiple cropped, c.1965

	Million hectares	Index of cropping
China	51.4	147
India	20.2	115
Bangladesh	3.4	139
Philippines	2.0	136
Egypt	1.8	173
Java	1.6	120
Japan	1.5	126
Pakistan	1.2	108
South Korea	1.2	153
North Vietnam	0.8	147
Burma	0.8	111

Source: D. Dalrymple, Survey of Multiple Cropping in Less Developed Nations, Foreign Economic Development Service, US Department of Agriculture, Washington, D.C., 1971

Multiple cropping

In most cool, temperate regions the length of the winter precludes the growth of more than one staple food crop, although several vegetable crops may be possible. However, in the subtropics and tropics temperatures are high enough for crop growth throughout the year provided moisture, from rainfall or irrigation, is available; hence two or even three cereal crops can be sown and harvested in a year. Multiple cropping has been practised in Egypt and parts of Asia for a very long period, and its extension is an important method of increasing the harvested area and total food output. It is difficult to establish with any accuracy, however, the proportion of the world's arable that is multicropped or the extent to which it has increased since 1950. D. G. Dalrymple's survey suggests that about 93 million hectares were multicropped in c.1965, 6.7 per cent of the world area then under arable (table 6.7). Most of the area multicropped was in Asia, and China and India had by far the largest areas. In Latin America some multiple cropping is found in the densely populated areas in the uplands of central America and the Andes; it was also practised in some of the irrigated areas of Mexico (but only 3 per cent of the total arable), the oases of the Peruvian coast and the interior of Argentina. In Africa multiple cropping is rare except in the irrigated areas of Egypt and the Sudan and some of the irrigation schemes of the savannah areas of West Africa.[7]

It is in Asia that multiple cropping is most common, although there are great variations in the intensity. This is normally measured by expressing the area harvested in a year as a percentage of the arable area, or of the net sown area (table 6.7). In Thailand only 1 per cent of the arable area is sown twice to crops in a year; in contrast the index of multiple cropping had reached 188 in Egypt in 1970–2 and 232 in parts of southern China, where three cereal crops can be grown in a year.[8]

Multiple cropping appears to have increased in Asia since 1950, although there are few reliable statistics. In India the index of cropping was 111 in 1947, but this had risen to 118.6 in 1970–1. In China there has been little increase in the net area in food crops since 1952, but the index of cropping rose from 131 in 1952 to 165 in 1980; increases in the index have also been recorded in Indonesia and Egypt. The intensity of multiple cropping is closely related to the density of population, and the expansion has been related to the growth of population and the need for extra food. However, its increase had been facilitated by the introduction of new varieties of wheat and rice that mature more rapidly than traditional varieties, thus making it possible to grow two cereal crops in a year. Although multiple cropping is possible in areas that rely upon rainfall alone, and is found in Java and the Philippines, it is generally closely associated with irrigated land, for whereas much of Asia has temperatures high enough for crop growth throughout the year, rather fewer areas have enough wet months to sustain two cereal crops.[9]

Irrigation and its expansion   Irrigation is by no means easy to define. It can encompass on the one hand massive reservoirs such as the Aswan dam in Egypt which stores the Nile's waters and releases them during the low season, and on the other simple earth dams in Sri Lanka that hold water from the monsoon rains which can be used in the dry season. Sometimes the term is used to describe the system of canals and ditches that carry the flood waters of a river to the distant parts of a flood plain.

The purpose of irrigation is not, of course, only to allow multiple cropping. Irrigation can be used to extend cultivation into areas where rainfall is too low for any crops to be grown, or to supplement rainfall, overcome variability and increase yields. In many humid areas such as England or the eastern United States high-value crops may be irrigated during dry spells. None the less, efficient irrigation systems do enable multiple cropping to be practised.

Although irrigation has been practised in Asia for several thousand years, much of the present irrigated area has been established only in the last ninety years.[10] The world irrigated area was only 8 million

Table 6.8 Irrigated areas of the world, 1950–88

	1950 (million hectares)	1988 (million hectares)	Increase, 1950–88 million hectares	per cent	Irrigated land as percentage of arable land, 1988
Developing					
Asia	83.5	142.8	59.3	71.0	31.6
Africa	5.8	11.1	5.3	91.3	6.0
Latin America	6.5	15.6	9.1	140.0	11.6
Total, developing	95.8	169.5	73.7	76.9	21.0
Developed					
North America	10.9	18.9	8.0	73.3	8.0
USSR	6.5	20.8	14.3	220.0	8.8
Australasia	0.7	2.1	1.4	200.0	4.2
Europe	6.4	17.3	10.9	170.3	11.7
Total, developed	24.5	59.1	34.6	141.2	8.6
World	120.4	228.6	108.2	87.3	15.4

Source: N. D. Gulhati, *Irrigation in the World: A Global Review*, New Delhi, 1955; FAO, *Production Yearbook 1989*, vol. 43, Rome, 1990

hectares in 1800, but reached 40 million in 1900, 120 million in 1950 and has since risen to 228 million; 15 per cent of the world's arable land is now irrigated. However, the significance of the irrigated areas is greater than this would suggest, for not only is multiple cropping possible, but also yields are higher and less variable than in rain-fed farming and high-value crops are grown. Irrigated areas thus account for a greater proportion of food output than the small area would suggest; one estimate is that 30 per cent of world food output comes from arable that is irrigated.[11]

Asia has nearly two-thirds of the world's irrigated area, and 32 per cent of its arable is irrigated (table 6.8). India and China have by far the largest areas of irrigation; they are followed by the United States and the Soviet Union, with 18 million and 20 million hectares respectively. In the developing world, most of the irrigated land lies in Asia. If Egypt and the Sudan are excluded, Africa has an insignificant area. In Latin America one-third of the total irrigated area is in Mexico, and nearly all of the rest is in Brazil, Argentina and Peru; the last has one-third of its total farmland under irrigation. Not only has Asia the largest area in irrigation but, of the developing regions, it has had by far the biggest absolute increase since 1950 (table 6.8).

Mixed cropping The term multiple cropping is normally used to describe the sequential growth of two or more crops in the same field in the same year. However, the growth of only one crop in a field or plot, although typical of Europe, North America and most parts of Asia, is less common in Africa and Latin America. In the *milpa* systems of Latin America and the bush fallowing of Africa plots of land are left fallow for several years between periods of cropping. But when the plots are cropped they are not necessarily sown to one crop; more commonly a mixture of crops is sown. The prime purpose of this mixed cropping is to protect the soil from the impact of raindrops and the direct rays of the sun and, if crops which require different periods to mature are grown, to space out the harvest period. Until recently most agronomists ignored mixed cropping, or assumed it was a primitive practice. But it has been shown that an area under several crops not only provides protection but also gives a higher output per unit area than the same area under one crop alone. First, if a variety of crops with different maturing periods is grown, then the whole year's available photosynthesis is utilized, not simply that of the period occupied by one crop. Second, different crops use different soil nutrients. Third, a mixture of crops prevents plant diseases establishing themselves. Mixed cropping is widely practised in Latin America and Africa; thus 60 per cent of Latin America's maize is grown with another crop, 40 per cent of the cassava and 80 per cent of the beans. In northern Nigeria 83 per cent of the arable area is in mixed crops. However, the extent to which mixed cropping has increased since 1950, and thus raised the calorific output of arable land, is unknown.[12]

The colonization of new land

Although there are no accurate figures on the increase in cropland due to the reduction of fallow or the spread of multiple cropping, some indication of the new land brought into cultivation since 1950 can be obtained by comparing the area in arable in 1950 and 1989 (table 5.1). In Latin America arable land increased by 94 million hectares or 109 per cent; in Asia by 103 million hectares or 30 per cent. The data for Africa are unreliable; it seems unlikely that there was an actual decline (see p. 75). In the 1970s new land in the world was being settled at the rate of 4–5 million hectares per annum, an addition to the arable area of 0.3 per cent per annum; but in the 1980s less than 3 million hectares were being added, at 0.2 per cent per annum.[13]

The reasons for land settlement have been various; increasing the food supply has not always been the prime motive. Thus in the aftermath of partition in 1947 the Indian government established colonies on waste land in India. Initially this was to settle refugees from Pakistan; since then the aim has been to settle landless labourers. This has been the purpose of land settlement schemes elsewhere, notably in Kenya and Java, both countries with a very rapid rural population increase. In some countries the landless were provided with land by expropriating part or all of the larger holdings, as in Iran, Taiwan, North Vietnam and China. Elsewhere land settlement has been seen as an alternative to land reform. In Latin America few governments have been prepared to change the inequitable distribution of land ownership, and the colonization of new land has been seen as a way of providing land for the landless without the problems of reform.

There are other reasons for land settlement. In south America the Amazon basin is divided among eleven countries; in each it is remote from the major centres of population, rural or urban, and sparsely populated. In the past settlements have been made for military reasons, to defend the land against the encroachments of neighbouring states. More recently the Amazon has been seen as an untapped area of major resources – not only food but also timber, oil and minerals – and nearly all the states have attempted, with varying degrees of success, to integrate the Amazon basin into their national economies. Military reasons led the Thai government to establish agricultural settlements on their borders with Laos, and in the Philippines the Huk rebels who surrendered to the government were settled on land in Mindanao.[14]

Nor have all land settlement schemes been primarily aimed at providing extra food. In Malaya, where schemes have been properly planned and efficiently executed compared with much of the rest of the world, much new land has been used for planting rubber and oil palms. Indeed the aim of many governments has been to prevent settlers reverting to subsistence; cash crops have been prescribed as part of the settlement policy.[15]

Land colonization schemes sponsored by governments and other organizations have had very mixed success. In India, although there has been a substantial increase in the area under crops, the cost to the state has been great and the return on investment very low. In terms of increasing the food supply the government of India would probably have been better advised to have invested in improving crop yields on the existing arable area. In Egypt 445,000 hectares have been reclaimed since 1952, and most of this was laid out in state farms. Yet crop yields on this land are only half those on the old lands, and on only one-third of the farms have incomes exceeded variable costs; i.e. two-thirds of the farms made a loss even before the cost of reclamation is considered. Nor have land settlement schemes made much impact upon landlessness. Thus in ten years Kenya settled 250,000 people, but this was no more than one year's rural natural increase. In twenty years 1 million people left Java under government sponsorship to settle in the outer provinces; in the same period Java's rural population increased by 20 million.[16]

New land in Latin America   In Latin America the expansion of the cropland has made a major contribution to the increase in food output; the harvested area rose by 70 per cent between 1950 and 1975, and two-thirds of the extra output came from increasing the area in crops. In Brazil in the same period four-fifths of the increased output came from extra cropland. This emphasis on increasing area rather than yields reflects not only the comparatively low population densities in much of the continent but also a long historical tradition.[17]

When the Spanish arrived in the Americas in the early sixteenth century they found that most of the lowland areas were sparsely populated by Amerindians who practised shifting agriculture or *milpa*. However, in the upland of Central America and the Andes there were more densely populated areas with sophisticated and intensive farming, although lacking the plough, the horse, cattle or sheep; the civilizations of the Incas and the Aztecs were the richest of these areas. It was here that the Spanish settled, for the control of labour was essential to mine silver or grow crops for export. The Portuguese settled the coasts of Brazil and imported African slaves to work sugar-cane plantations, and the Spanish later settled the land around the estuary of the Plata. But in the mid nineteenth century much of the continent was still very sparsely populated; there had been no substantial immigration as there had been in the United States. Nor had any class of independent farmer sprung up, and most of the land was held by a small minority. In the later nineteenth century, however, there was an expansion of the cultivated area, particularly in Argentina and to a lesser extent in Uruguay; Italian and other immigrants expanded the area in wheat and maize. In Brazil the coffee frontier expanded inland from São Paulo, and in the Caribbean and later the Pacific lowlands of Central America bananas or other tropical crops were raised on plantations for export.

The rapid population growth of this century has led to considerable demographic pressure in the upland areas, and there has been a slow move downwards from the uplands to the Caribbean and Pacific lowlands helped by the building of roads and the eradication of malaria. In Mexico not only was there a major land reform in the 1930s but the state invested in irrigation schemes, particularly in the arid north west. The area irrigated rose from 1 million hectares in 1926 to 4.3 million hectares in the 1960s, half of this on land not previously farmed.[18] But more dramatic has been the expansion of farmland in southern Brazil, south and west from São Paulo towards Paraguay and Uruguay, and northwards towards and beyond the new capital Brasilia. Between 1930 and 1970 Brazil's cultivated area more than quadrupled and Brazil became the world's second most important food exporter; since then the arable area has increased by a third. Much of this northward push was into the *cerrado*, an area similar in soils and climate to the African savannah and used until the 1970s only for extensive cattle raising.[19] But it is the Amazon basin, where comparatively little land has been brought into cultivation, to which most attention has been paid.

The Amazon basin　Until after the Second World War the Amazon basin was sparsely populated; it was lacking in roads and railways and river transport was poorly developed. In the late nineteenth century rubber collecting led to a boom around Manaus, and missionaries and soldiers had made their way down from the Andes to the *oriente* or *montana*, the forested slopes and lowlands to the east in the upper tributaries of the Amazon. This movement east from the Andes has been prompted by a number of factors. First has been the rapid growth of population in the upland areas of Colombia, Bolivia, Ecuador and Peru, which has led to both spontaneous and government-sponsored settlements in the upper Amazon basin. Second has been the discovery of oil – which attracted workmen who stayed to farm – and the building of roads. Third has been the military need to settle the land, for Ecuador, Colombia and Peru have all been in conflict over their boundaries. Between 1950 and 1975 85,000 people moved from the Bolivian *altiplano* to the *oriente*, 250,000 from the Ecuadorian uplands and some 100,000 from Peru.[20] In Brazil the building of roads – beginning with the Belém-Brasilia road in 1960, and accelerating in the 1970s with the construction of the TransAmazon highway – has led to a considerable movement into the Amazon. The population along the Bélem–Brasilia highway rose from 200,000 to 2 million between 1960 and 1970.[21] Initially the government of Brazil hoped the opening up of the Amazon would provide farms for the many landless from the drought ridden north east; but this has been far from successful. Hopes were of settling 100,000 in the 1970s. By 1977 only 6000 had moved from the north east into the Amazon basin. In 1975 the Brazilian government abandoned the welfare aspects of its Amazon programme and turned to development for profit, in alliance with large corporations, often foreign, to exploit not only the land but also timber and minerals.[22]

Since the 1960s much concern has been expressed about the future of the Amazon environment and the fate of the Amerindian population. Comparatively little rain forest remains in Africa or Asia, and the area in Amazonia constitutes one-half of the world total. But on this and many other matters it is hard to find accurate information for the basin as a whole. There is no reliable estimate of the proportion of the forest cleared, but it has been put as low as 5 per cent and as high as 15 per cent.[23] Although the use of remote sensing techniques in Brazil has increased knowledge of the basin, too little is still known of the soils and climate, although it is clear that there is far greater variety than once was thought. It used to be argued that only the *varzea* or alluvial soils of the river valleys were potential arable land; but they constitute only 5 per cent of the total area, and four-fifths are seasonally flooded. The *terra firme* or upland soils were once said to be uniformly poor, leached of their nutrients and liable to soil exhaustion once the forest had been cleared. That these soils are far more varied is undoubted, but there is as yet little knowledge of their value for long-term agriculture. Nor is there any proven system of permanent cultivation, other than growing tree crops or the extensive shifting cultivation and gathering practised by the indigenous peoples. However, experimental work suggests that sustained continuous cropping of annual crops is possible with the right combination of chemical fertilizers.[24]

The process of settlement has followed a similar course in both the Brazilian and the Andean Amazon. Although governments have encouraged the cultivation of cash crops, most settlers, whether spontaneous or in government schemes, have practised slash and burn, growing food crops such as rice, maize and yucca. Their holdings are generally small – less than fifty hectares – and many, after a few years of slash and burn agriculture, sow the land to grass and raise beef cattle, for which there is a market among the rich in both south-east Brazil and in the Andean cities. But the holdings of the original settlers are too small for extensive cattle raising, and they are frequently absorbed by large corporate ranches which use improved grasses, raise improved breeds of cattle and control plant and animal disease with pesticides. Some 80 per cent of the cultivated land in the Ecuadorian *oriente* is in grass, and elsewhere crops also seem to be a small proportion of the cleared area.[25]

Although there is no denying that there has been substantial settlement in the Amazon basin in the last thirty years, there is little reliable evidence on the areas cleared and settled, the area in crops or the quantities of food produced. Much that has been written on the subject has dealt with the threat to the Indian way of life or the problems that arise when the rain forest is destroyed. Yet it would seem that although some colonization has made a significant contribution to national food supplies – notably in Bolivia – on the whole the Amazon has not yet given a significant increase to Latin America's food supplies.

New land in Asia   Asia, like Europe, has a long history of settlement and by 1950 had very high population densities. Unlike Europe, most of the population were still employed in agriculture, and the density of the agricultural population was higher than elsewhere. Further, more of the total area was devoted to arable land – 16 per cent – than in Latin America or Africa. Much of the arable was in the deltas and lower alluvial plains of the great rivers, where wet rice cultivation has sustained high densities for two millenia or more. Before 1950 there had been a steady increase in the area in arable land over a long period, particularly in South-east Asia, where not only were more food crops grown but the development of plantations for the production of export crops became significant from the late nineteenth century.[26]

In the post-war period there has been both state organized and spontaneous colonization of land in nearly every country in Asia, in many cases prompted by the need to provide land for a rapidly growing landless population. In India there have been substantial additions to the cultivated area in a variety of regions, much of it by adding to the irrigated area which doubled between 1947 and 1973, particularly as the result of canal building and the construction of tube wells in the Punjab and Rajasthan.[27] In the north of the Ganges plain the *terai* or foothill area has been brought into cultivation since the eradication of malaria; that part of the *terai* which is in Nepal has seen a particularly rapid expansion of cropland since 1960.[28] In Assam, which was sparsely populated in the mid nineteenth century, migration and rapid population growth have converted the lowlands to a densely populated zone in the last half century. In Ceylon colonization schemes to settle the dry zone were begun at independence, and required the extension of irrigation. Some 70,000 hectares of paddy land were created in this manner between 1948 and 1964, and the total area of cropland increased by 40 per cent between 1948 and 1980.[29]

China, after India, has the largest area of arable land in Asia; in 1950 this was far more densely populated than the Indian subcontinent. More of the land was double cropped and irrigated, and yields were already high compared with much of the rest of Asia. Statistics of Chinese agriculture were published in the 1950s, but then little was known until the Chinese government released figures in the late 1970s. The old eighteen provinces of Han China were very densely populated, and there was little prospect of substantial expansion in this zone; some steep slopes in the south were terraced and attempts were made to convert cemeteries to cropland, and land on the coast north of the Yangtze and near the Hwangh-Ho delta were embanked and used for cropland. However, the Chinese looked to the north for their new lands. Manchuria had been closed to Han settlement by the Manchu dynasty until late in the nineteenth century; thereafter it became a classic zone of agricultural colonization until the Japanese invasion in the 1930s. The Chinese have certainly expanded cropland in this area, mainly by the establishment of state farms in unsettled areas, but the extent of this expansion is unknown.[30] Recent statistics suggest that although the sown area has increased since the 1950s, the arable area has declined. This may be due to soil erosion in the south and to urban expansion; but it may also be a result of underreporting of arable land by provinces in order to avoid taxation and the compulsory deliveries of food to the towns.[31]

In the archipelago of South-east Asia the settlement of some new land has required movement from island to island. At the beginning of this century the Philippines came under American rule: there was a marked contrast between the density of population on most of Luzon and some of the smaller Visayan islands, and the sparsely populated island of Mindanao and the north east of Luzon. Attempts to expand the area of settlement were begun by introducing legislation comparable to the American Homestead Acts, allowing the occupation of up to twenty-four hectares of public land. Movement to Mindanao became substantial in the late 1930s, with spontaneous movement exceeding assisted colonization. But by the 1960s population densities in Mindanao were approaching those in Luzon, and the very problems that colonization was supposed to overcome – fragmentation, tenancy and rural indebtedness – were appearing.[32]

In Indonesia the contrast between Java, with its very high population density, and the outer provinces of Borneo, Sumatra and the Celebes, with its very low population density, was and is very marked. In the nineteenth century Java under Dutch rule had experienced rapid population growth and by 1900 the centre and east of the island had very high rural population densities. The drainage of the northern plains had added to the arable area and elaborate systems of water control, multiple cropping and intensive cultivation had raised rice yields; in the east the adoption of cassava as a food crop had increased food output. However, there seemed little prospect of further increasing the cultivated area or of population growth slowing down.

Consequently in 1905 the Dutch introduced the policy of transmigration; Javanese were to be moved to the outer provinces, Sumatra in particular, both to relieve population density and to provide labour for the plantations on that island. Between 1905 and 1941 190,000 Javanese migrated, principally to Sumatra and mainly to the southern province of Lampung. Most of the migrants grew rice, as they had in their homeland. After independence the new government continued the policy of transmigration, and between 1952 and 1974 600,000 left Java. Further movements were planned in the late 1970s. This policy has no doubt increased Indonesia's cultivated area; whereas Java's arable area has remained stagnant since 1950, that in the outer provinces has increased considerably, but this has not increased the prosperity of the settlers or Indonesia's rice supplies. Nor have the numbers leaving Java been sufficient to improve the lot of the Javanese. In twenty years the total number leaving Java was no more than one year's natural increase in that island.[33]

Africa   Although African food output has increased since 1950, the rate of increase has been lower than in any other major region and has been below the rate of population increase for three decades. Although data on yields and land use are unreliable, most recent writers seem agreed that food crop yields have increased very little, and indeed may have declined. The bulk of increased output has thus been due to an increase in the area sown to crops. Between 1961–5 and 1976 the greatest increase in arable land was in East Africa, where it rose by one-third; the least was in the Sahel, where the increase was only 2.5 per cent and has since declined.[34]

In Africa multiple cropping has made little contribution to extra cropland, in contrast to Asia. Little of the continent's arable land is irrigated, and on that double cropping is rare, although the length of the growing season does not preclude it. Thus there are 70,000 hectares of irrigated land in the inland delta of Mali but only one-seventh is double cropped. This lack of double cropping does need some qualification. Multiple cropping is a term normally used to describe the sequential raising of one crop in a field. However, in much of Africa it is common to grow several crops on a plot; evidence suggests that this gives a higher dry matter yield per hectare than monoculture.[35]

The area under crops has thus been increased primarily by reducing the fallow periods and by the colonization of new land, either spontaneously or in government-sponsored schemes. In the last forty years there have been substantial increases in population density in nearly all parts of Africa, and this has led to changes in the type of farming. When densities were low, shifting agriculture could be practised. In this system forest or savannah land was cleared in small plots and a mixture of crops sown for two or three years. The plots were then abandoned

and trees and grass colonized the land; twenty years or more in natural
fallow was sufficient to restore soil fertility. Such a system was accompanied by a periodic movement of the village to be near the plots in
cultivation. The growth of population density has made such farming
impossible and it is now rare. Bush fallowing, with shorter periods of
fallow, remains predominant. The reduction of fallow has however
increased the frequency with which crops are grown and hence the area
in crops. But unless some means of maintaining soil fertility is introduced this reduction leads to a fall in soil fertility and yields and in
extreme cases to soil erosion, and the land may have to be abandoned.
Yields have fallen, for example, in Malawi, in the west highlands of
Kenya and in the Mossi plateau of Upper Volta.[36]

The spontaneous occupation of uncultivated areas has also occurred
in parts of Africa. This has been possible where the tsetse fly has been
eliminated and in such cases good land has been brought into cultivation. Elsewhere – as in parts of the Sahel, in Tanzania and in Ethiopia
– settlement has moved into semi-arid areas which have been liable to
crop failure from droughts and to soil erosion due to inadequate fallowing. There have been numerous land settlement schemes in Africa
both before and after independence. Increasing the food supply has not
been the only motive, they have not all led to an increase in cropping,
and most of the large schemes have been judged to be failures.

The first and still the largest settlement scheme in Africa began in the
Gezira in the Sudan before the First World War. At independence in
1955, 180,000 hectares were irrigated by the Nile, over half of which
were in cotton. Since then further irrigation has extended the area to
over 330,000 hectares. Other significant irrigation schemes were begun
by the French in Mali, Senegal and Mauritania and by the British in
northern Nigeria. The irrigation schemes in the West African savannah,
together with irrigation on European farms in Zimbabwe, made up
nearly all the African irrigated area outside Egypt, the Sudan and
Madagascar.[37]

Elsewhere in Africa land settlement schemes have had a variety of
motives. In Ghana and Zambia the building of dams for hydroelectricity works required the resettlement of large numbers living in the
flooded areas; in Kenya, Zambia and Zimbabwe Africans have been
settled on lands formerly owned by Europeans. In Zambia and Kenya
there have been schemes to resettle those living in densely populated
and eroded lands. Clearly not all land settlement schemes involved a
net increase in the arable area; it seems likely that in most parts of
Africa fallow reduction and spontaneous colonization – as in the
mechanized rain-fed agriculture of central Sudan – has contributed
more to the increase in food output than settlement schemes.[38]

THE LOSS OF LAND

There has been a substantial gain in the area sown to crops since 1950, although the precise figure is in doubt. There is no doubt again that some arable land has been lost, and in recent years much attention has been drawn to the loss of arable land in both the developed and the developing countries.

This concern is not new. In the 1930s the soil erosion in the west of the United States received much publicity and revived interest in soil conservation techniques. R. V. Jacks and R. O. Whyte pointed out that the loss of land through soil erosion and the fall in soil fertility threatened future agricultural output not only in the United States but in many other parts of the world, a theme that was taken up, in a more dramatic manner, by several writers in the late 1940s and early 1950s.[39] The fear that good agricultural land is being lost has been revived in the 1980s. In Britain there has been a long dispute about the loss of farmland to urban expansion and afforestation, and in the United States there have been similar controversies.[40] In the developing countries the loss of land to urban expansion has not been of great importance, except in countries such as Egypt. Two events have drawn attention again to the loss of farmland. The first has been the growing awareness that much of the world's tropical forests are being destroyed. The second was the drought in the Sahel in the years 1969–74 and over a much wider area of Africa in the 1980s; there was much debate about the causes of the famine at this time, and some argued that the Sahara was spreading southwards. Whatever the truth of this statement there has been a revived interest in the loss of farmland due to human mismanagement of the environment. Although much has been written about this, there is a singular lack of reliable data on the subject. Nor is this surprising, for the phenomenon is very difficult to define. There are, for example, parts of Africa where the reduction of fallow has led to gullying and the land has had to be abandoned: on the other hand over much larger areas fallow reduction has reduced the plant nutrient content of the soil and hence crop yields have fallen, but the land has not been abandoned. Thus there is a wide range of conditions contained within the terms *land degradation* or *desertification*. There has been a reduction of soil fertility and some land has been abandoned in many parts of the world, although there is no way of measuring the extent. Irrigated areas have presented particular problems. When large quantities of water are spread over farmland, much of the water is not used by crops but sinks downward, and eventually the permanent water table rises, causing waterlogging. Irrigated areas in hot regions are also prone to salinity. Water rises by capillary action and salts toxic to plants are deposited in the soil profile or on the surface. It has been claimed that one-fifth of the world's irrigated area is subject to either salinity or waterlogging and this has led to some land being abandoned and yields being reduced. The Near East is particularly susceptible to these problems. Half of the irrigated land in Syria is said to be affected by salinity or waterlogging, 30 per cent in Egypt, 15 per cent in Iran and, in the 1960s, three-quarters in Pakistan. These conditions are not irreversible. In Pakistan tube wells have been used to lower the level of the water table, and have reduced the extent of salinization.[41]

Soil is constantly being lost owing to natural processes. It is when loss exceeds the formation of new soil that problems arise. Once natural vegetation is removed and farming undertaken, soil loss is likely to occur in all regions. The risk of erosion is greatest in areas of low rainfall and very high temperatures and on steep slopes. Most traditional farmers have devised techniques to avoid soil erosion, but these may be ignored; in some cases their problems may be due to others' rather than their own activities. Thus in the Indus region of Pakistan flooding has been caused by the deforestation of the Himalayan foothills where the tributaries rise; this has accelerated run-off and caused flooding lower down. In Java recent deforestation of upland areas has caused the silting of irrigation systems in the lower parts of catchment areas. Where population pressure has increased the cultivation of very steep slopes, soil erosion has occurred, as in El Salvador and parts of Mexico. In Africa, particularly in parts of Kenya, Ethiopia and the western Sahel, population growth has prompted arable expansion into dry areas which are particularly prone to wind erosion. Nor is it cropping alone that causes soil erosion. The growth of cattle numbers in parts of tropical Africa and also in the arid regions of north-west Africa has caused overgrazing; as the number of palatable species declines, so have cattle numbers been increased, so that not only is the fodder supply reduced but land becomes progressively degraded. A similar sequence is occurring on the semi-arid pastures of northern Iraq, which are estimated to be able to support 250,000 sheep but in fact carry over 1 million.[42]

Examples of land degradation are legion, but figures on actual loss few; those that exist are often contradictory. It has been suggested that one-fifth of the current world arable area is subject to land degradation; and it has been argued that since the beginning of farming some 10,000 years ago an area equivalent to the present arable area has been lost to agriculture through soil erosion. The basis of this latter estimate is less than clear. None the less the loss of land is obviously significant; the growth of world food output since 1950 could have been greater without such losses.[43]

CONCLUSIONS

Although increases in crop yields have accounted for a high proportion of total world food output since 1950, increases in the area sown to crops have been important in the developing countries, particularly in Latin America and Africa. Increases in area have not come from the cultivation of new land alone. In Asia the increase in multiple cropping has been significant; in Africa the reduction of fallow has increased the area sown. The colonization of new land has also provided employment for the landless in the older established areas, although rarely for more than a small minority.

7

Agricultural Development in the Developed Countries since 1945

Few discussions of the world food problem deal with the growth of food output in the developed countries, where hunger is rare, nutritional problems are those of excess and not scarcity and agricultural problems are those of surplus not shortage. Yet there are good reasons for discussing food output in these countries.

First, malnutrition was far more common in the 1930s than it is now. In North America and Western Europe national food supplies were adequate and malnutrition was due to poverty and ignorance. On the other hand, in the Soviet Union and some countries in Eastern Europe food supplies were low and there was need for greater output, particularly of animal foods, to provide an adequate diet. Food supplies declined in the Second World War in the Soviet Union and most of Europe; during the war labour and other resources were taken out of agriculture, fighting disrupted food production and lack of transport made the movement of food difficult. Shortages continued for five or six years after the end of the war. In the late 1940s there were still fears about the future food supplies of Europe and Russia, although in the event recovery was rapid.

Second, although food output in Western Europe was restored to pre-war levels by 1950, and somewhat later in Eastern Europe, there has been need for an increase in food output since 1945 on two counts. One is that the population has increased by over 300 million in the developed countries since 1950 (table 4.1). The other is that rising incomes in the 1950s and 1960s increased demand for more livestock products. But, with the exception of the Soviet Union, consumption needs were largely satisfied by the early 1960s. Since then, however, not only has the population increased slowly, but rising incomes have been spent not on extra food but on other goods. Output, however, continued to rise and outpaced demand, so that for much of the late 1960s, 1970s and 1980s agriculture in Western Europe and North America has been plagued with food surpluses.

A third reason for considering production in the developed countries is the role of trade in the food supply of the developing countries. In the last fifteen years many developing countries have become dependent upon food imports in the form of either trade or aid. Although the cost of food surpluses in Western agriculture has been the subject of much criticism, the developed countries, and in particular the United States, have become the source of most of the world's grain exports. Without the production increases of the last forty years this would not have been possible.

THE GROWTH OF FOOD OUTPUT

Food output in Europe increased slowly in the first half of this century, but after 1950 grew at unprecedented rates. Food output in the developed countries approximately doubled between 1950 and 1980, although the rate of increase has slowed in the 1980s; output per caput, because of comparatively slow population growth, rose far more rapidly (see p. 82). There have, however, been considerable variations in the rate of increase between countries within the developed world. Some of the highest rates have been achieved in Eastern Europe and the USSR. Although productivity in Russian agriculture remains much lower than in the West, output increased very rapidly between 1950 and 1980; one estimate put it at 3.5 per cent per annum, over twice the rate in the United States.[1] In West Germany and the United Kingdom gross output doubled between 1950 and the 1970s; this was matched in several other countries in Western Europe. On the other hand, a few countries have had very little increase. Between 1952 and 1976 Norwegian agricultural output increased by only 5 per cent, and in Sweden, where government policy has been to restrict home output to 80 per cent of calorific needs, output actually declined in this period.[2]

Over most of the post-war period farming in Western Europe and North America has been heavily subsidized in various ways, which include tariffs, guaranteed prices and producer subsidies. This, combined with a remarkable technological advance has led to great increases in food output. But by the 1960s demand for food within Western Europe was beginning to slacken. Most Europeans were by then well fed, and increased prosperity did not lead to any increased expenditure upon food. Combined with a falling rate of population growth, this led to a very slowly increasing market for farmers. But because the European Community (EC) protected farmers from exports and provided guaranteed prices, output continued to increase. Since the 1970s output has increased at 2 per cent per annum in the EC, but demand by only 0.5 per cent per annum.[3]

By the early 1980s the cost of farm support was becoming politically unacceptable in the EC and various efforts – such as the introduction of quotas for milk production – were made to curtail output. In the United States the cost of farm support rose from $3 billion in 1981 to $26 billion in 1986, and as in Europe there were attempts to reduce output, which in both regions have been successful. In the 1980s food output in the EC has risen by only 7 per cent, and in the United States in 1989 was below the level of the early 1980s.[4]

THE EXPANSION OF THE ARABLE AREA

Traditionally the major means of raising food output has been to increase the area in cultivation, but, as has been seen (pp. 83–9), this has become less important as the capacity to increase crop yields has risen. In the post-war period most of the food output in the developed countries has come from higher yields; the arable area has shown little advance and indeed in many countries has fallen (see p. 76).

The area used for crops and grass has declined in nearly every country in Europe since 1950, although these *net* figures conceal more complex changes. In every country there have been losses due to urban expansion and, in some, to afforestation. This loss has been most serious in small and densely populated countries. In Belgium the agricultural area was two-thirds of the total land area in 1929, but in 1977 was only half. In The Netherlands reclamation of polders in the Zuider Zee has added to the agricultural area, but losses due to urban expansion have reduced the total cropland. In the 1950s, for example, some 3400 hectares were reclaimed, but 4000 were lost to urban expansion. In some countries losses in one region have been compensated for by gains elsewhere; thus in Canada marginal land has been abandoned in the east, particularly in Nova Scotia and New Brunswick, but there have been gains in the prairies.[5] In most of the developed world there have been few substantial additions to the arable area since 1950. There are, however, three exceptions to this.

First, in the Soviet Union, a remarkable expansion of the cultivated area was achieved in the 1950s. Some 36 million hectares of land were ploughed up in the semi-arid areas of Kazakhstan, the northern Caucasus and Siberia between 1954 and 1960, and Russian grain output was raised by 50 per cent. Since then, however, there has been little increase in the sown area, which was 202 million hectares in 1960 and 216 million in 1980. A large proportion of the present grain area is in these dry regions, and climatic fluctuations have led to great variations in Russian output over the last twenty-five years.[6]

Second, in the United States the *arable* area has not greatly changed. Losses of cropland from urban expansion in the north east have been made up by the reclamation of wet lands in the Mississippi valley and the irrigation of dry lands in the south west. The area actually sown to crops has, however, fluctuated with variations in United States government policy. At various times farmers have been paid not to sow land to crops and so the cropland has diminished; arable land has been left idle or summer fallowing has been increased. At other times there have been incentives to increase output and land withdrawn from cultivation at government behest has been sown again, and fallows in the semi-arid grain regions have been reduced to increase the sown area. Between 1945 and 1970 most of the increase in United States food output came from higher yields. Between 1970 and 1980, however, output rose by 20 per cent, yield by very little; much of this additional output came from an expansion of the sown area.[7]

A third region where area expansion has been important is Australia, where wheat accounts for a high proportion of the cropland. In the immediate post-war period increases both in the area in crops and in yields helped raise food output. But in the 1960s wheat cropping expanded into marginal lands where yields were lower than in the more humid regions. Since the mid 1960s the national wheat yield has stagnated, and area expansion accounts for the continued increase in crop output.[8]

Although increased area has played an important role in increasing food output in some regions of the developed world, its importance has been dwarfed by increases in the yield of crops and animals. Before considering these improvements, however, some structural changes in agriculture must be touched upon.

STRUCTURAL CHANGE IN THE DEVELOPED COUNTRIES

In the developing world the agricultural labour force doubled between 1950 and 1989 and thus has contributed, by more intensive cultivation, to increased food output. Quite different circumstances obtain in the developed world. The agricultural labour force was in slow decline in

Table 7.1   Labour force in agriculture, developed countries, 1950–80

	United States	Western Europe	Eastern Europe and USSR
1950	23,048	42,096	76,842
1960	15,635	33,146	67,616
1970	9,712	22,626	49,363
1980	6,051	16,445	38,622

Sources: United Nations, Economic Bulletin for Europe, **35** (2), 1983, p. 172; G. C. Fite, American Farmers: The New Minority, Bloomington, Ind., 1981, p. 101

the United States and north-west Europe before 1945, but was still increasing in Russia, Australasia, Eastern and southern Europe. Since 1950 there has been an unprecedented decline in the workforce in the developed countries, which in 1980 was less than half what it was in 1950 (table 7.1). There are various reasons for this decline, but of greatest importance has been the attraction of much higher wages in the urban economy, so that in the period of remarkable economic expansion between 1945 and 1973 many millions left the land for the towns. This has led to the substitution of labour by machinery and power, which has occurred, at varying rates, throughout the developed world. In every country the tractor has replaced the horse, and tractors have become bigger and faster; the combine harvester, hardly known in Europe before 1945, now harvests nearly all grain. A variety of machines unknown or in the experimental stage before the war now harvest crops which were once hand picked, such as tomatoes, tobacco, cotton, potatoes and sugar-beet. Livestock production has been similarly transformed. In 1950 only 3 per cent of the cows in the EC region were milked by machine, whereas by 1975 only 3 per cent were milked by hand; in addition, haymaking has been mechanized.[9] Not only has the horse been replaced as a source of power, but so too has man.

In the 1930s few farms in the United States had electricity supplies, although about half the farms in Western Europe were connected. Now the main tasks in the farmyard and farm buildings, once done by hand, are performed with the aid of electricity. The growth in the use of machinery and the dramatic decline in the labour force has led to remarkable increases in labour productivity. In 1950 in Western Europe it took a hundred man-hours to cultivate 1 hectare of cereals; by 1980 it took only ten man-hours. In the United States agricultural output rose by 55 per cent between 1950 and 1972, but in the same period the number of million man-hours worked in a year on the land fell from 15,137 to 6172.[10]

Table 7.2 The decline in farm numbers, 1960–83

	Number of farms[a] (thousands)		1983 as a percentage of 1960
	1960	1983	
United States	3963	2428	61
Belgium	209	90	43
Denmark	193	97	50
France	1774	1061	60
West Germany	1385	744	54
Italy	2756	1927	70
Ireland	278	221	79
Netherlands	230	124	54
United Kingdom	443	244	55

[a] Farms over 1 hectare.
Sources: G. C. Fite, *American farmers: the new minority*, Bloomington, Ind., 1981, p. 101; D. R. Bergman, 'The transition to an overproducing agricultural system in Europe: an economic and institutional analysis', in G. Antonelli and A. Qiadiro-Curzio (eds), *The agro-technological system toward 2000: a European perspective*, 1988, pp. 54–74

Not only farm labourers but farmers have left the land in North America and Europe, particularly the occupiers of very small farms (table 7.2). This has allowed other farms to get bigger and has made the use of machines economically possible. In the United States the average size of a farm rose from 80 hectares in 1950 to 100 hectares in 1980. In Europe average sizes were, and remain, much smaller, but have still shown considerable increases. In Belgium, for example, the average size in 1950 was 5 hectares but was 11 hectares by 1980; and in Sweden the average arable holding rose from 12.5 hectares in 1951 to 20.1 hectares in 1971, and in Denmark from 16 hectares to 24 hectares between 1960 and 1977.[11]

SCIENCE, TECHNOLOGY AND GOVERNMENT SUPPORT

The primary consequence of mechanization has been to greatly increase labour productivity and cut increases in the cost of production. It has also allowed better and faster cultivation and more timely sowing and harvesting, thus increasing yields. But the increases in yields have been mainly due to other causes.

During the first half of this century crop yields rose slowly in Western Europe and even more slowly in North America and Australia, where there was an abundance of land, and farmers aimed at maximizing output per caput rather than output per hectare (see figure 5.4). In the 1930s most farmers in Europe still relied upon growing crops in rotation and using farmyard manure to maintain soil fertility and to increase crop yields, although some chemical fertilizers were applied, particularly in the Low Countries and West Germany. In Britain only potatoes and sugar-beet received any significant amounts. Plant breeding institutions were established in several countries after the rediscovery of Mendel's theory in 1900; the exchange of new varieties also led to some increases in yields. The Danish barley Kenia, for example, was adopted in England and France. Perhaps the most significant plant breeding advance at this time was the development of hybrid corn in the United States. First bred in 1914, it began to be grown commercially in the early 1930s. Weeds were controlled by rotations, by deep cultivation with a mouldboard plough and by hand hoeing. Some inorganic pesticides were known but were little used, and effective selective herbicides or fungicides were not available. Improvements in livestock breeding and feeding were spreading. Artificial insemination was first used in Denmark in the 1930s and ideas on balanced feeds for livestock had been developed in Germany and were being adopted elsewhere; in Britain, The Netherlands and New Zealand there was much research on grass varieties and grassland management. But in the 1920s and 1930s farming in most of the developed world was depressed and few farmers could afford to innovate, either in yield increasing improvements or in purchasing machines such as tractors, combine harvesters or milkers.[12]

However, the emergencies of the Second World War and the continued food shortages immediately afterwards led nearly all governments to encourage greater agricultural production, and a variety of guaranteed prices, deficiency payments, subsidies and tariffs brought prosperity back to farming; farmers began to adopt the improved technologies made available in the 1920s and 1930s. Since then most states have encouraged research in all branches of agriculture, and a continuous flow of new machines, crop varieties, fertilizers and pesticides has provided farmers with the means of yet further increasing yields and output.

HIGHER CROP YIELDS

The application of new techniques in farming has raised yields at unprecedented rates since the end of the Second World War. Thus it has

Table 7.3   Wheat and milk yields in Western Europe

	Milk			Wheat		
	1948–50 (kg/cow)	1988–9 (kg/cow)	Increase (%)	1948–50 (kg/ha)	1988–9 (kg/ha)	Increase (%)
Netherlands	3657	5876	60	3530	7412	110
Belgium	3367	4192	25	3143	6702	113
Denmark	3117	6207	99	3587	6987	95
Sweden	2703	6170	128	2223	5683	155
United Kingdom	2673	4801	80	2687	6386	137
Switzerland	2843	4818	70	2647	6340	140
West Germany	2173	4782	120	2470	6522	164
Ireland	1933	3960	105	2243	7884	251
France	1730	2915	68	1830	6249	241
Norway	1963	5846	198	2190	3832	75
USA	–	6447	–	1120	2247	100
Australia	–	3858	–	1120	1594	42
USSR	–	2558	–	840	1828	118

Sources: United Nations, *Economic Bulletin for Europe*, **35** (2), 1983, pp. 167, 168; FAO, *Production Yearbook 1989*, vol. 43, Rome, 1990; *Production Yearbook 1964*, vol. 18, Rome, 1965

been estimated that in Western Europe wheat yields increased at 0.8 per cent per annum between 1850 and 1913 and at 0.6 per cent between 1913 and 1938, but at 2.2 per cent, between 1939 and 1970 and reaching 2.6 per cent in the 1960s. Wheat yields have doubled and in some cases tripled since the 1950s (table 7.3); even in those countries where yields were already high in 1950, such as The Netherlands and Belgium, there have been considerable gains, although the land abundant countries such as the United States and Australia still have much lower yields than north-west Europe. Most arable crops have had substantial increases in yields, but generally increases have been greatest in the cereals; increases in sugar-beet in Britain (table 7.4), for example, have been much less than those in wheat, potatoes and barley, and in the United States soybeans and potatoes have lagged behind maize and wheat (table 7.4).

The precise reasons for the increase in yields are difficult to determine because there have been so many changes in technology in the last thirty years. New varieties have been adopted, more chemical fertilizers have been used, pesticides and herbicides have reduced crop damage and the competition of weeds, irrigation and underdrainage have improved environmental conditions, whilst the use of machinery

Table 7.4 Crop yields in the United States and United Kingdom (tonnes per hectare)

	United States				United Kingdom		
	1950	1989	Percentage increase		1950	1989	Percentage increase
Wheat	1.05	2.2	110	Wheat	2.7	6.6	144
Maize	2.5	7.0	180	Barley	2.5	4.8	92
Potatoes	19.2	32.2	68	Potatoes	18.2	35.8	97
Soybeans	1.6	2.2	38	Sugar-beet	26.6	40.4	52

Sources: Ministry of Agriculture, *Agricultural Statistics for the United Kingdom,
1980 and 1981*, London, 1982; *Century of Agricultural Statistics in Great Britain,
1866–1966*, London, 1968

has allowed sowing and harvesting to take place at optimum times. However, research in both Britain and the United States suggests that the breeding of new cereal varieties has been the single most important cause of increased yields, accounting for approximately half the recorded gains.[13]

Plant breeders have been able to improve cereals in a variety of ways. The breeding of varieties with a high response to fertilizer has been important, and the introduction of short stemmed varieties, able to carry a larger head of grain, has been crucial. Immunity to specific diseases has been significant, and the traditional breeders' art of producing varieties for different environments has continued. In the United States hybrid corn varieties were originally grown in the corn belt. Local varieties were necessary when the hybrid was extended north into Ontario, south west into Texas and south east into the old cotton belt. In Britain Scandinavian barleys have been crossed with indigenous varieties to produce a variety that can be grown in wetter and cooler regions, helping to explain the remarkable expansion of barley growing in Britain since the 1930s.[14]

Plant breeders have produced a succession of improvements. Farmers have adopted new varieties rapidly, not only within countries but across international borders. Thus the French wheat Capelle-Desprez was introduced into Britain in 1954; by 1965 88 per cent of the winter wheat area was sown with it. But the subsequent introduction of yet better varieties led to its equally rapid demise; by 1975 only 3 per cent was still sown with the variety. The Maris winter wheat, which replaced Capelle-Desprez, was equally rapidly adopted in France. The speed with which new varieties are introduced must not, however, be exaggerated.

In Britain there is a lag of some twelve years between the planning of a new variety and the distribution of new seed to farmers.[15] The majority of discussions of yield increases place most emphasis upon the rapid growth of fertilizer consumption since 1950. Certainly the growth cannot be denied; the consumption of fertilizers tripled in the United States between 1950 and 1972, rose tenfold in Canada between 1939 and 1975 and increased fivefold in the countries of the EC between 1950 and the 1970s. This, however, possibly overstates the gain in plant nutrients in the soil; prior to the Second World War chemical fertilizers were comparatively expensive and large applications of farmyard manure, which contains nitrogen, phosphorus and potash, the essential nutrients, were made by the better farmers. Chemical fertilizers were also applied in the intensive farming areas of north-west Europe. Since then chemical fertilizers have greatly increased in consumption but in parts of Europe and the United States less farmyard manure is now applied.[16]

In the 1950s and 1960s the use of nitrogen, phosphorus and potash fertilizers increased at much the same rate, but by the 1960s the phosphorus and potassium content of soils in much of Western Europe could be maintained with relatively light applications of these nutrients. Hence much of the subsequent increase has been in straight nitrogen, which is the most common limiting factor in increasing crop yields (figure 7.1). By the 1970s fertilizer consumption in Belgium, Norway, The Netherlands and West Germany was thought to be near the optimum, as it was on much of the cereal crop in England. But comparatively little was used on grassland; whereas leys receive fertilizer, permanent grass receives little.[17] There are, however, still considerable variations in fertilizer consumption within the developed world. Fertilizer applications are far less in the extensively farmed areas of North America and the Soviet Union than in the intensively farmed region of north-west Europe (table 7.5).

Although new varieties and the increase in the supply of plant nutrients have been of critical importance in raising crop yields, the adoption of chemical herbicides, pesticides and fungicides has also made a valuable contribution. Before 1945 weeds – any plants which compete with the crop for plant nutrients in the soil – were kept down mainly by cultivation with a mouldboard plough, which inverts the soil and destroys weeds, and by hoeing during the growth of the crop. The invention of selective herbicides that destroyed weeds before or during growth has reduced labour needs, and has both allowed cultivation with tines and some direct drilling and helped to increase crop yields. Chemical pesticides and fungicides were known before 1945, but were not efficient and were little used. Discoveries during and since the war

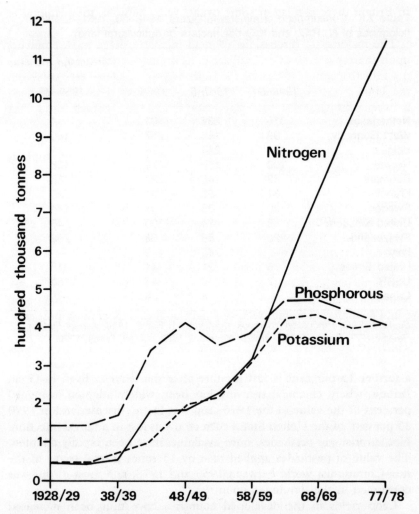

Figure 7.1 Use of nitrogen, phosphorus and potassium fertilizers in the United Kingdom, 1950–80 *Source*: Sir Hans Kornberg, *Agriculture and Pollution*, Royal Commission on Environmental Pollution, Seventh Report, London, HMSO, Cmnd 7644, 1979

of a number of efficient pesticides have been followed by much research in the chemical industries. In Britain some 800 different chemical sprays have been marketed, and the appropriate machinery has been developed and is now widely used. The importance of chemical spraying should not be underestimated. The total loss to world food output from pests, diseases and weeds has been put at one-third of the standing crop, and

Table 7.5 Consumption of mineral fertilizers, 1948–50, 1961–5, 1988 (kilograms of N, P_2O_5 and K_2O per hectare of agricultural land)

				Percentage increase
	1948–50	1961–5	1988	1950–88
Netherlands	315	232	301	−4
West Germany	99	185	257	160
Belgium	161	244	275	71
Norway	108	154	225	108
Denmark	82	146	220	168
France	31	82	194	525
Sweden	39	83	107	174
United Kingdom	88	74	130	48
Switzerland	25	56	88	252
Italy	–	45	123	–
United States	8	21	41	412
USSR	2	6	45	2150
Canada	–	8	28	–

Sources: FAO, *Production Yearbook 1957*, vol. 11, Rome, 1958; *Annual Fertilizer Review 1977*, Rome, 1978; *Fertilizer Yearbook 1989*, vol. 39, Rome, 1990

a further 15 per cent is lost in store after the harvest. Even in Great Britain, where chemical spraying has been widely adopted, some 10 per cent of the value of the 1976 crop was lost to disease, and in 1970 15 per cent of the United States corn crop was lost to a fungal infection. Not surprisingly pesticides, once available, have been rapidly adopted. The value of pesticides applied rose by 15 times in real terms in the non-Communist world between 1953 and 1978, and most of this was consumed in the developed countries.[18]

Crop yields in the developed countries have thus been increased because farmers have purchased extra inputs developed and made off the farm. New plant varieties have been bred by state breeding institutes and seed merchants and fertilizers, pesticides and herbicides produced by the chemical industries. Changes on the farm have also improved the crop's environment. Irrigation is not of major importance in the developed countries, but its extension has helped increase crop yields as well as allowing dry lands to be cultivated. In the United States the irrigated area has tripled since 1940 and now produces one-quarter of the total value of agricultural output. In much of northern Europe poor drainage once limited crop yields but the underdrainage of heavy soils has reduced waterlogging and helped increase yields.[19]

Table 7.6   Livestock output in the United Kingdom (thousand tonnes)

	1930s	1978
Beef and veal	550	1048
Mutton and lamb	190	238
Pork	210	634
Poultry meat	89	726
Milk (million litres)	8.4	14.9

Source: W. Holmes, 'Animal husbandry, 1931–1980', in G. W. Cooke (ed.),
*Agricultural Research 1931–1981: A History of the Agricultural Research Council
and a Review of Development in Agricultural Science during the Last Fifty Years*,
Agricultural Research Council, London, 1981

Table 7.7   Livestock slaughtered in Canada (thousands)

	Cattle	Pigs	Sheep and lambs
1951	1734	4488	438
1961	2731	5850	633
1971	3251	9743	205
1976	4331	7493	188

Source: M. Troughton, *Canadian Agriculture*, Budapest, 1982, p. 166

LIVESTOCK PRODUCTION

Although most discussions of post-war agriculture deal with advances
in the productivity of arable land, livestock production makes up a
majority of agricultural output; it accounts for at least half the total
value of agricultural output in every country in the developed world,
varying from 55 per cent in France and the USSR to over 85 per cent
in Denmark. Over the last forty years animal production has generally
increased more rapidly than crop production, the acceleration of a process that began about a century ago in some Western countries, more
recently in others.[20]

In nearly all Western countries the total output of animal products
has increased substantially since 1950; Canada and the United Kingdom illustrate widespread tendencies (tables 7.6 and 7.7). These increases in output have been achieved partly by increases in the number
of animals kept, partly by increases in their productivity.

All livestock numbers have increased in the major regions of the

Table 7.8 Increase in livestock numbers in the developed countries, 1950–80 (thousands)

	1948–52	1987	Percentage change, 1950–87
Cattle			
Western Europe	66,335	94,324	42
North America	88,369	112,402	27
Australia	19,458	30,738	58
USSR	55,780	121,348	117
Sheep			
Western Europe	46,510	100,685	116
North America	32,000	11,253	−65
Australia	145,000	221,855	53
USSR	76,800	141,490	84
Pigs			
Western Europe	34,778	123,092	254
North America	63,626	57,673	−9
Australia	1,702	3,092	82
USSR	19,720	78,452	298
Chickens			
Western Europe	369,519	872,500	50
North America	508,044	1,449,500	185
Australia	19,188	65,000	238
USSR	452,000[a]	1,127,000	149

[a] 1958–9.

Sources: FAO, *Production Yearbook 1963*, vol. 18, Rome, 1964; *Production Yearbook 1988*, vol. 42, Rome, 1989

developed world since 1950, with the exception of a decline in sheep and pigs in North America (table 7.8). As there has been little change in the agricultural area over this period, livestock densities have risen considerably; this has only been possible as a result of changes in the methods of feeding livestock. Animals can be fed in a variety of ways, but the fundamental contrast is exemplified by the differences between Ireland and Denmark or the United States and Australia. In Denmark 90 per cent of the farmland is in crops most of which are fed to cows and pigs. In Ireland 90 per cent of the farmland is in grass and this provides much of the diet of livestock. In the United States 90 per cent of arable output is fed to livestock, in Australia only 10 per cent. Thus in Ireland and Australia ruminants – cattle and sheep – rely primarily on grass for their fodder, in Denmark they rely upon arable crops.[21]

In the 1930s there was a greater reliance upon grass in all developed countries; but in several countries in Western Europe considerable areas were devoted to fodder crops of which potatoes and fodder roots were among the most important.

Some grain was fed to cattle, and concentrates, made from imported oilseeds, were also used. Pigs and poultry, which are non-ruminants and cannot easily digest grass and therefore have to be fed with crops that can also be directly fed to man, were fed with a variety of feeds including grain, potatoes, fodder roots and skimmed milk.

In the last forty years there have been substantial changes in the nature of livestock feeding and the efficiency of fodder production. Grass remains a major source of feed in some countries, notably Ireland, Britain, New Zealand and The Netherlands, but the output per hectare has increased substantially. Research on grass varieties has led to the introduction of higher yielding varieties; in Britain, for example, Belgian and Dutch varieties have been adopted. There have been considerable increases in the application of fertilizers, particularly nitrogen fertilizers, and the use of lime – notably in Ireland – has reduced soil acidity and increased yields. The introduction of electric fencing has improved the management of grazing lands, and the cutting of hay has been mechanized, whilst the spread of silage making has increased the efficiency of grass preservation.[22]

Perhaps greater changes have taken place in the use of arable crops as fodder. Root crops, which were an important part of the fodder supply in the 1930s, needed much labour to weed and harvest, and in most countries have declined as labour has become scarcer. Some new fodder crops have been introduced; soybeans and hybrid grain sorghum have been most important in the United States, and in western Europe maize, a high yielding fodder, has advanced northwards. The principal change has been the increased use of cereals, and particularly barley, as a feed for cattle, pigs and poultry. By the early 1980s nearly three quarters of the grain consumed in developed countries was fed to animals; but fewer crops grown on the farm – with the exception of grass – are now fed *directly* to livestock. Knowledge of the dietary needs of animals at different stages of their growth has led to the idea of balanced rations, and much livestock feed is now bought from firms who produce compounds made up of grain and high protein foods; the latter were formerly provided in Europe by imported African oilseeds or fishmeal, but these have been increasingly replaced in the last ten years by soybeans imported from the United States and Brazil.[23]

There has thus been a considerable growth in the supply of fodder for livestock, partly from higher yields and partly, in Europe and the USSR, by increased imports of feed. But improvements in livestock and their environment have increased the efficiency with which fodder is converted into milk, meat and eggs. Since the 1930s advances in the knowledge of livestock diseases, and the development of vaccines and antibiotics, has greatly reduced the impact of disease. Among cattle the elimination of bovine tuberculosis and the great reduction of brucellosis have been of critical importance, and there have been parallel advances in the reduction of disease among pigs and poultry. Indeed, the confinement of large numbers of pigs and poultry in small areas would have been impossible without control of disease. Cattle have been bred to improve the efficiency with which feed is converted into meat. In Britain milk formerly came mainly from Friesians. West European breeds such as the Charollais, Simmental and Limousin have become established in North America and Britain. Modern pigs and poultry are largely hybrids, replacing the local varieties. Indeed modern broiler hens in Western Europe are mainly of American origin. The control of the productive capacity of livestock has been increased by the spread of artificial insemination, which began in Denmark, Britain and the United States in the 1930s but is now a more widespread practice.[24]

Better quality animals, which are now better fed and better housed, have greatly increased livestock productivity. The increase of milk yields is one index of this (table 7.3); but there are other measures. The output of eggs per hen doubled in Britain between 1960 and 1980, the number of piglets per sow per annum rose from eleven in 1946 to seventeen in 1976, and the average number of lambs per ewe has increased. For these and other reasons, the efficiency with which fodder is converted into meat or milk has increased, most noticeably among pigs and poultry. In 1946 in Britain, it took 5 kilograms of feed to produce 1 kilogram of pig meat, but by the 1970s only 3.5 kilograms were needed. For poultry the feed needed to produce 1 kilograms of meat fell from 3 kilograms to 2 kilograms over the same period.[25]

There have been other changes in the efficiency of livestock production in the last thirty years. It is perhaps necessary to note some changes in the organization of livestock raising. In the 1930s there were some important regional differences in the means of producing meat and milk, but in Europe pigs and poultry were generally produced on mixed farms, which also raised crops and other livestock. Indeed poultry were a very subsidiary part of mixed farming, and only in Denmark and the corn belt of the United States was pig production specialized and intensive. Since 1945 pig production and poultry rearing have in many countries become separated from general farming; special buildings and controlled lighting, ventilation systems and the purchase of all feed – so that little land is needed – has produced a system of meat production which is very different from traditional farming practices and, although efficient, has alarmed many critics of factory farming. These systems have been part of a general tendency for much larger herds and flocks in livestock production. Indeed the changes in livestock husbandry since 1945 have been as radical as those in arable farming and the advances in output per man as dramatic as those in arable production.[26]

CONCLUSIONS

During the last forty years agriculture has undergone a revolution more profound than anything experienced in the past. Although there remain marked differences in productivity between countries within the developed world, there have been marked increases in output and productivity over time in all countries, whether in the East or the West. By the late 1950s agriculture in the developed countries had fully recovered from the war and was providing a diet which met the demands of an increasingly affluent public. Since then output has continued to rise owing to further increases in productivity. Scientific advances made in government and some private research institutes have provided new ways of reducing the time spent on farm operations, increasing crop yields and output per animal. But by the 1960s most of the population of the developed countries were able to buy a satisfactory diet with a large proportion of livestock products. Further increases in incomes did not lead to yet further increases in the consumption of foodstuffs, and problems of surplus emerged in many countries in Western Europe and North America.

Throughout the period most countries, for political and strategic reasons, have protected their farmers against imports by tariffs, by guaranteed prices or by subsidized inputs. Methods of protection have varied but they have generally been such as to provide small and inefficient farmers with a reasonable living. But technological advance has had serious consequences. Pre-war farmers bought relatively few inputs; most were produced on the farm. Over much of the last thirty years input prices have risen more than agricultural prices, so that costs have risen more than incomes; farm real incomes compared with nonfarm incomes, have declined. Thus, in spite of the great increases in food output, neither consumer nor farmer have been satisfied.

8
Tropical Africa

Since the end of the Second World War no part of the developing world has had such a sad agricultural history as the countries of tropical Africa. In 1945 all but Ethiopia and Liberia were European colonies. Ghana was the first country to become independent, in 1957, and the other British and the French colonies soon received their freedom; colonial power finally ended when Mozambique became independent in 1974. Economic changes have been less fortunate. Between 1945 and 1960 most parts of Africa prospered. The economic recovery of Europe provided a rising market for Africa's agricultural products, and for much of this period the terms of trade favoured primary exporters. Food output at least kept pace with population growth, and indeed may have exceeded it; food imports in the 1950s were very small.[1] But since 1960 the situation has deteriorated. Food output has of course increased, but the rate of increase is now below the peak rate of increase achieved in the 1960s; and in each decade since 1950 has been below other developing regions.

This has been compounded by changes in population growth. Whereas in every other region the rate of increase in the 1980s was lower than in previous decades, in Africa the rate has risen continuously since 1945 and in the 1980s was in many countries growing at rates rarely experienced elsewhere in the world. Between 1980 and 1987 the average rate of increase in Kenya and the Ivory Coast exceeded 4.0 per cent per annum and in six other countries it was above 3.5 per cent. As a result food production per capita in sub-Saharan Africa has declined continuously since the 1960s, and by the late 1980s was lower than in any year since 1960 (figure 8.1). Food *supplies* per capita have also fallen, but not as greatly as food production, because of the growth of imports, principally of cereals. In the early 1960s only 1 million tonnes were imported, by the mid 1980s this had risen to over 12 million tonnes. The burden of these imports is greater than appears, for the cost rose by 50 per cent in the 1960s and sixfold in the 1970s. These imports

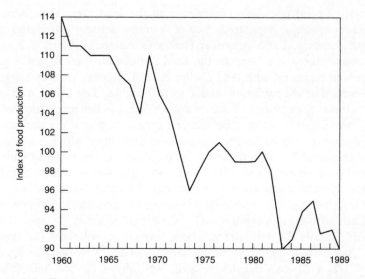

Figure 8.1 Food production per caput in sub-Saharan Africa, 1960–89; 1976–8 = 100 *Sources*: C. Christensen, 'Food security in sub-Saharan Africa', in W. L. Hollist and F. L. Tullis (eds), *Pursuing Food Security – Strategies and Obstacles in Africa, Latin America and the Middle East*, London, 1987, p. 68; FAO, *Production Yearbook 1989*, vol. 43, Rome, 1990

partly reflect an increased demand for wheat in African towns, where incomes are higher; little of the imports make their way to the rural areas. None the less Africa's self-sufficiency in food has fallen from a ratio of 98 per cent in 1962–4 to 90 per cent in 1972–4 and 79 per cent in 1983–5.[2]

The problems of African farmers were compounded by the difficulties of their economies as a whole in the 1970s. In the 1970s gross national product (GNP) per capita fell – if Nigeria is excluded – by 0.5 per cent per annum, and current account deficits rose from $1.5 billion to $8 billion in 1980.[3] In the 1980s the loans contracted in the 1970s caused a major economic crisis. The burden of debt and interest rose to enormous proportions. By 1987 simply paying interest took half the value of the region's exports; official development assistance declined and private lending almost ceased entirely, whilst GNP per capita fell throughout the 1980s, as did the value of agricultural exports.[4]

POPULATION AND LAND

Africa is frequently described as sparsely populated. It is also often described as a continent with an abundance of farming land. This view requires qualification, on a number of grounds.

First, if total population is related to the land area, then Africa is certainly sparsely populated; but it is more densely populated than North America, Latin America, Russia or Australia (figure 8.2(a)). If agricultural area is related to the total population then Africa appears to be well endowed with land (figure 8.2(b)). Further enquiry suggests, however, that the situation is not so favourable. The urban and non-agricultural populations of Africa are a much smaller proportion of the total population than in other major regions, so that if the agricultural population is related to the agricultural area then Africa's density is only exceeded by Asia (figure 8.2(c)). Much of the total agricultural area is poor grazing land – the semi-arid regions of the Sahel and East Africa. If only arable land is considered, then the density of the agricultural population to arable land is seen to be considerably higher than in Latin America, although well below the densities of Asia (figure 8.2(d)). If only the agricultural labour force is considered – the agricultural population includes the dependants of those working in agriculture – then Africa's density is again well above all but Asia (Figure 8.2(e)). Finally, if the agricultural population is related to the area in the major food crops then tropical Africa's density is only a quarter less than the average for Asia, twice that of Latin America and far above any of the developed regions (figure 8.2(f)).

Second, the *distribution* of population must be considered. Whereas there may well be considerable areas in Africa that are underpopulated, there are certainly regions of high density, and a considerable proportion of the population is concentrated in a small area. In about 1970 one-third of Africa's population lived on only 2 per cent of the total land area.[5] The major centres of population in tropical Africa were (figure 8.3) the coastal areas of Benin and Nigeria with some 50 million people, Hausaland in northern Nigeria with 25 million, Rwanda-Burundi and areas around Lake Victoria with 25 million, and central Ethiopia with 15 million, all living at high rural densities. The concentration of population is particularly marked in East Africa where half the total population of Tanzania, Uganda and Kenya live in the Kenya highlands or near Lake Victoria on only 13 per cent of the total area.[6] Clearly, although much of Africa has very low population densities it is not without problems of population pressure.[7]

Third, population density must be related to the carrying capacity of the land. It is widely agreed that it is difficult to measure carrying capacity. However, a group of agricultural experts has recently attempted to relate the existing population to the potential food output.[8] They have done this by estimating the length of the growing season and calculating the calorific output of the most productive food crops. They then calculated the numbers that could be supported for some 10,000 unit

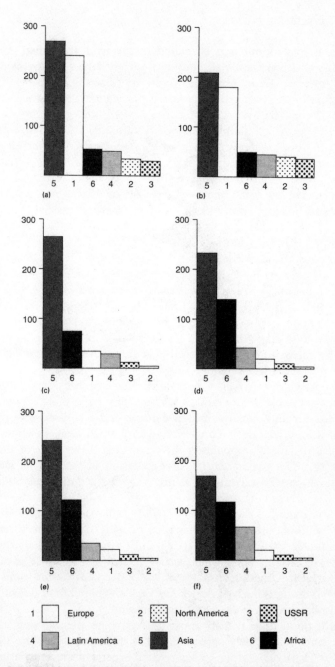

Figure 8.2 Total and agricultural population densities in the major regions, 1988 (world average equals 100): (a) total population per land area; (b) total population per agricultural area; (c) agricultural population per agricultural area; (d) agricultural population per arable area; (e) agricultural labour force per arable area; (f) agricultural population per area in major food crops *Source*: FAO, *Production Yearbook 1989*, vol. 43, Rome, 1990

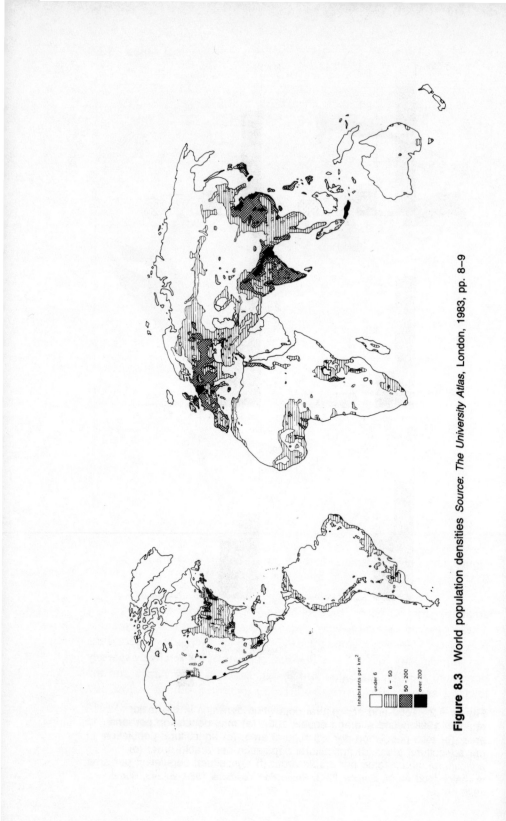

Figure 8.3 World population densities *Source: The University Atlas*, London, 1983, pp. 8–9

Inhabitants per km²

under 6
6 – 50
50 – 200
over 200

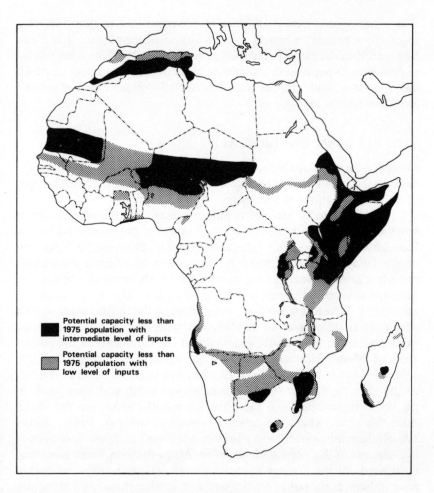

Potential capacity less than
1975 population with
intermediate level of inputs

Potential capacity less than
1975 population with
low level of inputs

Figure 8.4 Carrying capacity and population density in Africa *Source*:
G. M. Higgins et al., 'Africa's agricultural potential', *Ceres*, **14**, 1981, pp. 13–21

areas in Africa at three levels of input – low, intermediate and high.
These carrying capacities were then compared with the actual population in 1975. In two regions the 1975 population already exceeded the
carrying capacity (figure 8.4). The first consists of the relatively sparsely
populated areas of the Sahel and Sudan zones of West Africa and the
lowland areas of East Africa, where the potential output is low and
much of the area is devoted to pastoral activities or rain-fed agriculture
of a low productivity. The second consists of the high density areas in
the relatively fertile districts of Rwanda-Burundi, the eastern shores of
Lake Victoria and the Kenyan highlands.

It is not argued here that Africa is overpopulated or that high densities are a primary obstacle to agricultural development. It is clear, however, that the rapid population growth of the last fifty years has led to problems of population pressure in parts of Africa, that not all Africa has abundant land supplies and that not all African farming systems have low labour inputs.

THE AFRICAN ENVIRONMENT

European views of the African environment have ranged from the excessively optimistic to the unnecessarily pessimistic; some brief account of its salient features is necessary.

Low temperatures are not a limiting factor in the distribution of crops or livestock in tropical Africa. In some of the upland regions of East Africa cold does limit cultivation but this affects only 0.3 per cent of the land of the continent.[9] In most parts of Africa temperatures provide a growing season long enough for a wide range of tropical and subtropical crops, although it is not easy to grow the major temperate zone crops. Wheat, for example, is confined to the upland areas of East Africa. It is differences in rainfall, not temperature, that determine variations in the agricultural geography of the continent. In the coastal areas of West Africa and in the Congo basin annual rainfall is high, generally over 1400 mm. This region has two seasonal peaks of rainfall, as the intertropical convergence zone moves north and then south in the northern summer, and there are few months which are dry. Away from this zone, where the original vegetation was rain forest, annual rainfall diminishes and the dry season lengthens; there are corresponding changes in the vegetation. In West Africa the rain forest gives way northwards to the Guinea zone, or humid savannah, then the Sudan zone, where grass rather than woodland predominates and there are five to seven dry months and then the Sahel where grass and thorn bush predominate and the dry season may be as long as ten months of the year.[10] Beyond lies the Sahara. In East Africa the arid zone occupies much of the Horn and the lowland areas of Kenya and Tanzania. South of the Congo basin savannah areas of varying types occur and, although annual rainfall may be high, dry seasons of varying intensity predominate.

It is the length of the dry season that is critical for farming in much of Africa; nor is this merely a matter of the rainfall received. The high temperatures that prevail over much of tropical Africa throughout the year give very high rates of evaporation – four times those in the British Isles – and reduce the moisture available for crops.[11] The length of the growing season varies considerably; in the humid tropics there is little or no dry season and perennial crops can be grown. But away from the equatorial zone annual rainfall declines and so does the length of the growing season, and only annual crops are possible. With diminishing rainfall, variability increases; the amount of rainfall received varies considerably from year to year, as does the timing of the onset of the rainy season, and this, in the savannah regions, has a profound influence upon crop yields. Only half of Africa has sufficient rainfall for rain-fed agriculture and, in much of this zone, yields fluctuate considerably from year to year.[12]

African soils have been a matter of much debate and are still not properly understood. Compared with many other parts of the world Africa has disadvantages. Much of the continent has been without earth movements for very long geological periods, so there are few recent deposits to form the parent material of soils. Recent earth movements have given volcanic activity only in limited areas in Cameroon and East Africa. There are no large deposits of boulder clay, as in North America or parts of Europe. Furthermore, during the long periods of geological stability the upper layers of deposits have been subject to chemical decomposition and leaching. Finally, for a combination of physical and cultural reasons the deltas and lower reaches of African rivers – with the exception of the Nile – have not proved suitable for agriculture, in marked contrast with Asia.[13] The very high temperatures, and in the forest regions the high rainfall, have created problems for farmers. Under a forest cover soil temperatures rarely exceed 26 °C, but when the vegetation is removed temperatures on the soil surface may reach 42 °C, accelerating the rate of plant decomposition and of leaching. Most of the plant nutrients in the forest ecosystem are stored in the vegetation, and its removal to allow cultivation leads to a rapid decline in the nutrients and humus in the soil, to adverse changes in soil porosity and to an increase in acidity. The most successful farming systems in the forest are those that simulate the forest cover, either by planting perennial shrubs and trees or by growing an intermixture of cereals, roots and shrubs that provide protection from the high temperature and the impact of rain drops. It does not follow that African soils are inherently infertile or that they cannot be farmed successfully; it is simply that the traditional methods of farming, with long fallows of natural vegetation to restore fertility, have been undermined by the growth of population, and no entirely satisfactory replacement has yet been devised.[14]

MAJOR TYPES OF FARMING IN AFRICA

African farming shows a great deal of diversity. Some of its characteristics are outlined briefly in the following section.

Crops and livestock

A wide range of crops is grown in Africa. In the forest regions perennial crops such as coffee and rubber can be grown; food crops include maize and bananas, but the distinctive features of these regions are the root crops – yams, taro and manioc, the last also being found in drier regions. West and central Africa is the only part of the world where root crops, with their low protein content, form a large part of the calorie intake. Rice is not of importance except in Madagascar, in Sierra Leone, where it has long been grown as an upland crop, and in small areas of irrigation in the West African savannah. In the savannah regions perennial crops cannot be grown without irrigation. The dominant food crops are a variety of millets and sorghum in both West Africa and eastern and southern Africa, where, however, maize is also widely grown. Groundnuts and cotton are the characteristic cash crops of the savannah areas.[15]

Most African farmers produce only crops; livestock are kept by pastoralists. Few farmers use ox drawn ploughs (see p. 155) and the integration of crop and livestock production is rare. Much of Africa is infested with the tsetse fly, which transfers trypanosomiasis, a disease endemic in African game, from the game to cattle. Cattle other than the indigenous dwarf varieties are not numerous in most of humid tropical Africa; they are found mainly in the drier regions which lack the thick bush which provide the tsetse fly with its natural habitat. Approximately one-third of Africa is affected by the tsetse fly. Its eradication, it has been calculated, would allow an increase of some 125 million cattle.[16]

Population, soils and fallowing

Perhaps the most distinctive features of African farming are the methods of maintaining soil fertility and the relationship between population density and land use.

At the beginning of this century population densities throughout most of Africa were low and the most common form of farming was shifting cultivation. When land was to be planted with crops the natural or secondary vegetation, whether forest or savannah, was cleared with the aid of an axe or machete; the larger trees were left and the stumps of others often not cleared. The vegetation was burnt. Crops were then planted with the aid of a digging stick; there was little or no preparation of a seed bed, although in the forest regions some root crops were planted in mounds of soil. Unlike in Europe or Asia the plot was planted not with one crop but a mixture, which might include roots, cereals and perhaps shrubs. Little or no weeding was practised, no fertilizer or cattle manure was applied and after two or three years the land was abandoned and the natural vegetation allowed to recolonize the plot. Other plots were cleared for cultivation, and in some cases the village huts moved to the new site. Twenty or thirty years in natural fallow restored plant nutrients to the soil and the land could be sown to crops again. Such a farming system was almost universally condemned by Europeans. However, by the 1930s it began to be realized that this farming system, in Lord Hailey's words, was 'less a device of barbarism than a concession to the character of the soil'.[17] Later, Allan argued that the length of the fallow and the techniques employed were a function of inherent fertility: the poorer the soil, the longer the fallow needed. Agronomists showed that fallows, if of sufficient length, were capable of maintaining soil fertility in the long run, and that many of the features of the system were an excellent adjustment to the tropical environment.[18] Thus, leaving stumps in the cleared plots accelerated later colonizations of the fallow. Burning the vegetation made the soil friable and the ash added some plant nutrients to the soil. Growing a mixture of crops, particularly a combination of roots, cereals and shrubs, provided cover for the soil, reduced soil temperatures and protected the soil from the impact of rain drops. More recent research suggests that the growth of a combination of crops can give higher output per unit area than single stands.[19] As noted earlier, a fallow of adequate length does restore soil fertility. Many would argue that shifting agriculture, or bush fallowing as this method is now generally called, although giving low yields gives quite high outputs per head, for, in the absence of much soil preparation or weeding, labour inputs are low. A further advantage is that growing a combination of crops with different periods of maturing staggers the harvest dates.

More recently a different interpretation has been put upon the development of bush fallowing. Shifting cultivation requires a very long fallow, and thus needs abundant land. As population increases, less land is available and the fallow period is reduced. This requires some alternative means of maintaining crop yields such as weeding, the use of manures, and the planting of fallows with rapid growing grasses or leguminous plants, all this requiring extra labour.[20] Thus the major variations in farming in tropical Africa have been related to increasing population density; as density rises so the fallow is reduced, practices that help maintain crop yields are adopted and farming moves from extensive to intensive methods.[21] If, however, alternative means of maintaining soil fertility are not adopted then yields fall and soil erosion occurs.[22]

African farming then can be differentiated upon the basis of the

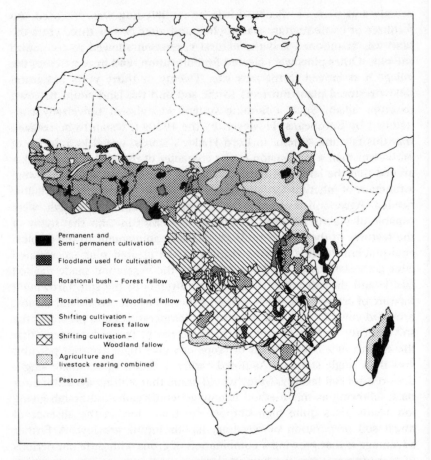

Figure 8.5 Types of farming in Africa *Source*: W. B. Morgan, 'Peasant agriculture in tropical Africa', in M. F. Thomas and G. W. Whittington (eds), *Environment and Land Use in Africa*, London, 1967, pp. 242–73

length of fallow, which in turn may be related to population density (figure 8.5). Shifting cultivation, where the site of the village is moved and fallows exceed twenty years, is now less common. In West Africa it is confined to a few parts of the forest. In central Africa – the Congo basin – and in much of southern Africa it is still the dominant mode, owing partly to the lower population densities but also to the relative absence of towns compared with West Africa. But the growth of population density has reduced its extent. In Kenya, for example, it was common in the 1930s but is now rare. Bush fallowing, where the fallows are shorter and neither forest nor woodland becomes re-established, occupies most of West Africa and the Sudan zone that stretches into Ethiopia. Probably three-quarters of Africa's cultivated land is in either shifting cultivation or bush fallowing. Permanent agriculture, where fallows are non-existent or very short, occupies little area but supports much of the population. It occurs over most of the Kenyan highlands, eastern Madagascar where wet rice is grown, the Hausa area of northern Nigeria, south-eastern Nigeria and parts of Burundi and Rwanda.[23]

The problems of bush fallowing are twofold. First, where adequate measures to maintain soil fertility are not taken, soil fertility is not restored, crop yields decline and soil erosion may occur. The first signs of this problem came in the 1930s. Not only was population beginning to increase after a long period of stability, but European occupation of land, notably in Kenya and southern Rhodesia, had limited the fallow area available to African farmers. In the 1930s and 1940s colonial governments attempted to introduce conservation methods, with limited success. Since 1950 the population of Africa has nearly tripled, and the agricultural population has more than doubled. Throughout Africa the length of the fallow has been reduced and this was frequently followed by declining yields and in some cases soil erosion. Although yields of food crops in Africa are not known with confidence there is evidence of decline in areas where fallows have been reduced – in the groundnut zone of Senegal, in Niger, in the Mossi plateau of upper Volta, the highlands of west Kenya and in Mali, Sudan and Ghana.[24]

Second, most recent technical advances in agriculture have been made in temperate areas and are adapted to the climatic and technological conditions of permanent agriculture; their introduction into Africa requires considerable modifications of bush fallowing. Thus the use of the plough or tractor to increase the area in cultivation requires the complete clearance of vegetation and makes natural fallows difficult to restore. Although African soils are low in plant nutrients, particularly phosphorus, the use of chemical fertilizers is difficult, for high temperatures and heavy rainfall in the forest zone leach them downwards out of the soil; in the drier areas they cannot be absorbed by the crop. The cultivation of a variety of crops in cleared plots precludes the use of pesticides, which are specific to a single crop, and harvesting machinery cannot be used except on single stands. The introduction of large scale farming using the plough and other heavy machinery has led to some notable disasters in colonial times and since.

Small farms and commercialization

There are large farms in Africa. Thus in Zimbabwe and Zambia European farmers with large holdings still exist, although in dwindling numbers. In Kenya many European farms that passed to Kenyans after independence did not change in size. Plantations are to be found, although few in both numbers and as a proportion of total farmland. In some African countries large-scale farming on the Soviet model has been tried, notably in Ghana, and in the Sudan large-scale mechanized farming of grain has developed in the last thirty years. But the great majority of farms in Africa are small. Thus in Ethiopia the average size of a farm is less than 5 hectares; in Tanzania 83 per cent of farms are less than 3 hectares, in Malawi its 80 per cent. In the Kano close-settled zone the average family holding is over 3 hectares, but only 1.2–1.5 hectares are in crops; the rest is in fallow. Farms in southern Nigeria are also overwhelmingly small.[25]

But these small farms provide most of Africa's food output and a substantial proportion of its export crops. Forty years ago African farming was described as predominantly subsistence; it is still true that a majority of African farmers produce food primarily for their own consumption. It has been estimated that only half of the African total agricultural output is marketed, and of this, half is export crops, half food crops; a more recent estimate suggests that only 15–25 per cent of *food* output is sold off the farm. The growth of a large urban market, combined with the existence of a substantial rural population without land – 8–10 million rural Africans are estimated to have no land – has led to an increase in the production of food crops for cash rather than simply for farm consumption. Most African farmers still produce their own food as a first priority but many also grow cash crops. Only in a few areas, however, have farmers given up food farming to concentrate upon cash crops, as in the groundnut areas of Senegal, in some of the cocoa areas of Nigeria and in the cotton areas of Gezira in the Sudan.[26]

Few economists now argue that African farmers are unresponsive to price changes, or that their farming behaviour is determined primarily by cultural rather than economic norms. None the less, African governments have had much difficulty in stimulating peasant farm output and in securing food supplies from the countryside for the towns.

THE GROWTH OF AGRICULTURAL OUTPUT

Export crops

In the second half of the nineteenth century the growth of incomes and industry in Europe provided Africa with the opportunity to export a variety of agricultural commodities, notably oilseeds. Initially European involvement was solely through traders, but by the early twentieth century most of tropical Africa had become French, British, German or Portuguese colonies. The rapid growth of agricultural exports after 1900 came from three types of farms. First, Europeans settled in Kenya, southern and northern Rhodesia and the Portuguese colonies of Mozambique and Angola. Exports from European farms included tobacco, tea and coffee and in some colonies Africans were forbidden to grow these cash crops in order to maintain prices for European settlers. In northern Rhodesia much of the maize for consumption in the copper belt was produced on European farms. Second, land was leased to foreign companies to produce crops on plantations. Thus in the Congo land was leased to Unilever to produce palm oil, sisal was grown in Tanganyika, rubber in Liberia and sugar in Mozambique. In West Africa, however, both the British and French colonial governments were reluctant to allow either plantation companies or European settlement. Third, African smallholders adopted cash crops which were sold to European traders in the major ports. The most notable examples of indigenous production were cocoa in Ghana and southern Nigeria, groundnuts in Senegal and northern Nigeria and cotton in Uganda.[27]

The growth of these export crops was very rapid between 1900 and 1930; Nigerian exports, for example, rose fivefold, then declined during the depression of the 1930s and during the Second World War and then boomed again in the immediate post-war period.[28] The extent of the area under cash crops in about 1950 is not known with any accuracy, but it has been estimated that it occupied 15 per cent of the area cultivated by the indigenous population,[29] and in the first half of the century the area under export crops grew more rapidly than the area in food crops.[30] Since the 1950s there have been changes in both the organization and the production of export crops. In Kenya African smallholders have rapidly adopted crops that they were formerly forbidden to grow, such as tea. Coffee has emerged as a major export crop, particularly in the Ivory Coast and Kenya. In some areas plantations have been nationalized – notably in Tanzania and the former Portuguese colonies – and the number of European settlers has declined in Kenya, Zambia, Angola and Mozambique. Elsewhere, on the other hand, state plantations have been encouraged in the newly independent countries. More to the point, the great boom in the volume of agricultural exports levelled off in the mid 1960s and in the 1970s earnings declined in many countries; in the 1980s African agricultural exports declined at 1.8 per cent per annum. There have been a number of reasons for this. First, the prolonged drought of the 1970s and early 1980s reduced output in the savannah zones. Second, the rise in petroleum prices in the 1970s meant that imports from the developed countries became expensive relative to African exports, and after 1979 the terms of trade for African exports became even more unfavourable. Third, the recession in the industrial countries in the late 1970s and early 1980s reduced demand in the industrial countries for Africa's exports. For some exports, such as coffee and tea, the income elasticity of demand was very low; the market was saturated, whilst soft drinks and artificial chocolate were substituted for imports. Fourth, for some exports other developing countries have emerged as more efficient producers, notably in palm oils but in other commodities as well. In the 1970s some well-established exports declined substantially, such as cocoa from Ghana, and in the late 1970s Nigeria was importing groundnuts and palm oil.[31]

The role of exports in African economic development has been subject to much debate. Twenty-five years ago René Dumont commented on the absurdity of groundnuts being exported from countries where humans had too little protein to be fed to cattle in Europe.[32] Others have argued that the substantial area devoted to export crops – possibly 10–15 per cent of the harvested area in tropical Africa[33] – should be used to grow food crops for home consumption. But African states are here in a difficult position, for exports of agricultural products account for a high proportion of foreign earnings in all except Nigeria and Zambia, where oil and copper provide most export earnings. Loss of these earnings would greatly retard African countries' attempts to modernize their economies and improve their standards of living.

Food Crops

Little is known of the growth of food output before 1950; since then it has increased, but at a lower rate than in Latin America or Asia; indeed in 1972–4 and 1983–5 output fell in a number of countries. But in all three decades since 1960 the population of sub-Saharan Africa has increased more rapidly than food output, so that output per capita has fallen continuously (figure 8.1). In the 1960s this fall only occurred in the arid zone of the Sahel and in parts of southern Africa (figure 8.6), but in the 1970s there were only five countries in the continent where output per capita did not decline, and in the 1980s only seven. The lack of reliable statistics on food crops and the confusion between fallow and cultivated area make it difficult to estimate the relative importance of area increase and higher yields in the growth of food output in Africa. Most authorities believe that yields have increased little and extra output has come mainly from increasing the area in crops. Thus the area in crops is thought to have increased by 12 per cent between 1960 and 1975, with the greatest increases in East Africa but an actual decline in the Sahel. During the same period 90 per cent

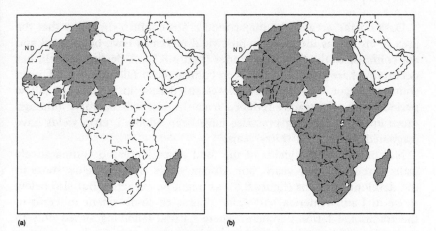

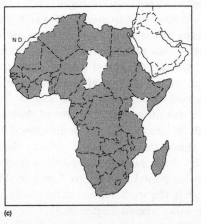

Figure 8.6 Countries where food output per caput declined: (a) 1961–5 to 1969–70; (b) 1969–70 to 1979–80; (c) 1979–80 to 1988–9 *Source*: FAO, *Production Yearbook*, Rome, various dates

of the increased cereal output came from area expansion, only 10 per cent from yield increases. However, yield increases have become more significant, so that between 1960 and 1990 half the increase in cereal output came from higher yields; none the less half came from expansion of area, a greater proportion than in any other region. Increases in area have come not only from the expansion of settlement into un-occupied areas, particularly in the Sudan, but from the reduction of fallow and an increase in the period for which crops are grown between fallows. Expansion into new areas has often been into marginal areas, as in the Sahel during the rainier periods of the 1960s, or into the drier regions of northern Buganda in the 1950s.[34]

Data on the area in the major food crops confirm the belief that the area in crops has increased, whilst yield increases have been small and are confined to the last two decades (figure 8.7). A decline in yield has occurred where poor land has been cultivated, or fallows reduced, as in southern Niger.[35] In the Kano close-settled zone in Nigeria the fallow period has been steadily reduced over the last half century; although more intensive farming practices have been adopted, crop yields have stagnated for the last thirty years.[36]

Not only have the yields of the food crops increased comparatively little in the last forty years, but African yields are well below those in the developed regions (figure 8.8), as might be expected, but also below those in Latin America and Asia. This is to some extent inherent in the traditional farming system, where natural fallowing aimed only to maintain yields and where labour inputs were minimized so that although yields were low, output per head was high. Once fallow was reduced, some means of maintaining – let alone increasing – yields was necessary. The traditional separation of crop production and livestock keeping precluded the use of animal manure, and the absence of tropical legumes as effective as those of temperate regions has excluded another way of increasing nitrogen in the soil.

In the developed countries and in parts of Asia and Latin America traditional methods of farming had been developed by this century that gave crop yields above those found in Africa; they have since been further increased by the application of modern methods of farming. In Africa this has not happened. Inputs are lower in Africa than in either Latin America or Asia (figure 8.9); not surprisingly, crop yields remain lower than in any other region (figure 8.8).

THE USE OF CHEMICAL FERTILIZERS

Chemical fertilizers were almost unknown in Africa in 1950, and were confined to plantations and some of the cash crops of European farmers, such as tobacco in southern Rhodesia.[37] But the use of chemical fertilizers was equally limited in Latin America and Asia forty years ago. African consumption of fertilizers has certainly increased since then, but the level of utilization is much lower than in Asia or Latin America (table 8.1). Furthermore, the use of fertilizers has remained concentrated on relatively few farms and generally upon cash crops. Thus in Kenya, one of the more successful African states in post-independent times, nearly half of all fertilizer consumption in the late 1970s was for three cash crops, coffee, tea and sugar-cane; only 3 per cent was used by smallholders for food crops.[38] There are numerous reasons for

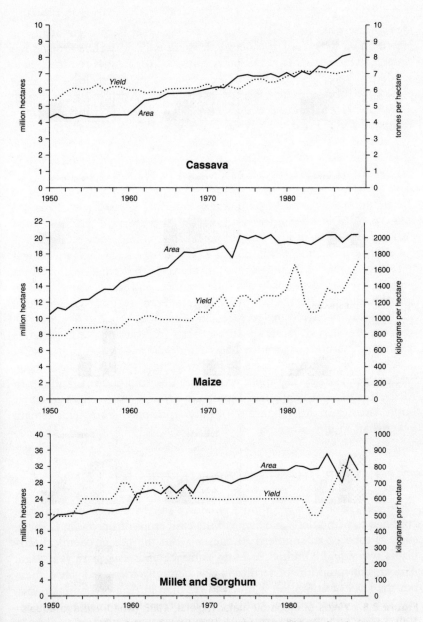

Figure 8.7 Trends in the yield and area of major African food crops, 1960–80 *Source*: FAO, *Production Yearbooks*, Rome, various dates

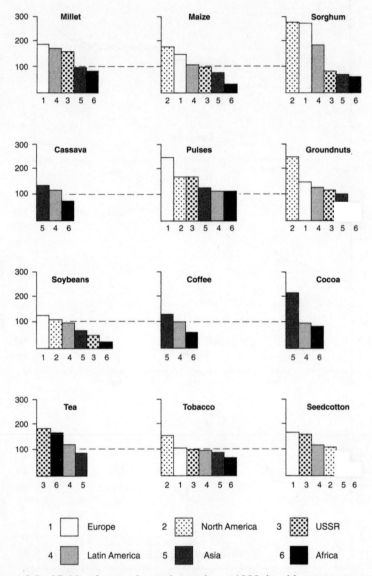

Figure 8.8 Yields of crops by major regions, 1988 (world average equals 100) *Source*: FAO, *Production Yearbook 1989*, vol. 43, Rome, 1990

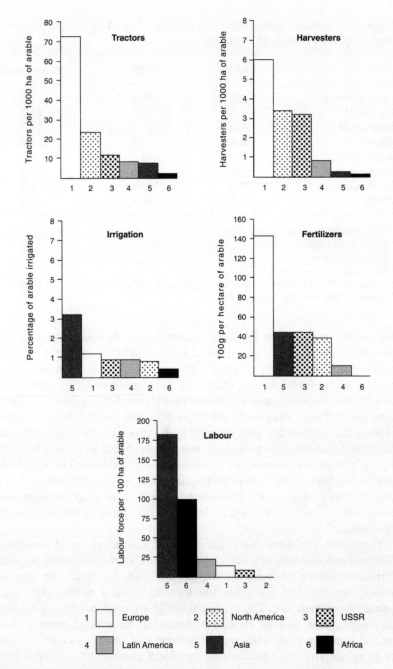

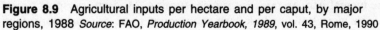

Figure 8.9 Agricultural inputs per hectare and per caput, by major regions, 1988 *Source*: FAO, *Production Yearbook, 1989*, vol. 43, Rome, 1990

Table 8.1 Fertilizer consumption from 1949–51 to 1988 (kilograms per hectare of arable land, all nutrients)

	1949–51	1988	Increase factor
Latin America	3.1	49.0	15.8
Far East[a]	1.6	70.9	44.3
Near East[b]	2.4	61.7	25.7
Africa[c]	0.4	10.8	27.0
Asian CPEs	–	242.6	–
Developing	1.4	76.8	54.9
Developed	22.3	124.6	5.6
World	12.4	98.7	8.0

CPEs, centrally planned economies.
[a] Excluding China and Japan.
[b] Excluding Israel.
[c] Excluding South Africa.
Sources: FAO, The State of Food and Agriculture 1970, Rome, 1971; Fertilizer Yearbook 1989, vol. 39, Rome, 1990

the limited increase in fertilizer consumption. Although African soils are deficient in plant nutrients and crops might be expected to show good responses to fertilizer, there are agronomic problems in optimizing its use. In the forest regions where the forest is cleared the combination of very high temperatures and heavy rainfall leaches the nutrients down into the soil and away from the roots of crops. In the drier savannah regions the unreliability and, in much of the savannah, the lower amounts of rainfall mean that fertilizers are not dissolved and so cannot be taken up by plants, but instead are oxidized on the surface. The successful use of fertilizer in these regions requires the use of irrigation, which is uncommon (see p. 154). Moreover, most African crop varieties have been selected not to respond to fertilizer but rather to survive drought or disease.[39]

Economic problems have also delayed the adoption of fertilizers. Where land remains abundant there is little economic point in using fertilizers; where densities are higher, and only fertilizers can increase yields, farmers find the price too high in relation to the extra yield gained from their use. The chemical fertilizers used in Africa are nearly all produced in Europe or North America; the cost of moving them inland from the ports adds greatly to their cost and, in a country where tractors are still few, distributing fertilizers around the scattered plots of the typical African farm is time consuming. Modern transport routes in Africa were built to link cash crop zones to ports, and many food farmers are still dependent upon head loading to market their crops and to acquire inputs. Thus, in Ghana and Sierra Leone some 70 per cent of produce is still moved to market by head load. For many, if not a majority, of African food producers, the use of fertilizers is economically impossible.[40]

CROP VARIETIES

Much of the increase in crop yields in Europe and Asia since 1950 has been attributable to the breeding of crop varieties that are immune to disease and responsive to fertilizer use. However, the breeding of new varieties is specific to the climate and soils of a region, and although the technique of breeding can be adopted in other climates, the seeds themselves can only rarely be directly transferred. In Africa the little research on agriculture in colonial times was directed to improvements in export crops. Since 1950 great advances have been made in the Americas and Asia in breeding new high yielding varieties of maize, wheat and rice. But for the most part these advances have not been adopted in Africa. Rice is little grown in Africa; methods of rice growing comparable with those in Asia are found in eastern Madagascar, but otherwise rice growing was confined to Sierra Leone where indigenous rices were grown using shifting cultivation. Since the 1930s wet rice has been produced there and also in some irrigated schemes in the West African savannah. But the successful cultivation of high yielding rices needs careful water control and irrigation, which is uncommon in Africa. Furthermore it has proved difficult to transfer high yielding Asian rices; of some 2000 varieties tried in West Africa, only two yielded higher than local varieties. New wheat varieties have been successfully adopted in Latin America and Asia, but wheat can only be grown in the upland areas of East Africa; new varieties have been adopted with success in parts of Ethiopia.[41] Hybrid maize, originally bred in the United States and then successfully adopted in Mexico, has proved more successful in Africa, where, unlike wheat or rice, it is a major food crop. Maize varieties bred in Kenya and issued in 1964 were rapidly adopted by smallholders; by 1973 half Kenya's maize was sown with hybrid seed, giving yields considerably above that of traditional seed, whilst by the mid 1980s 85 per cent of smallholders in Zimbabwe were using improved varieties. Successful adoption has also occurred in Zambia and parts of Ethiopia, but only a small minority of all African farmers have so far adopted improved varieties, not least because its successful growth requires heavy fertilizer application. Unfortunately comparatively little progress has been made in improving the other African staples of roots, millets and sorghum.[42]

IRRIGATION

In much of the African savannahs rainfall is not only low but is concentrated in a short period, limiting the types of crop that can be grown. The onset of the rainy season, to which period cultivation is confined, is variable, and the lower the rainfall the greater the variability. There are thus large parts of Africa where crop cultivation is impossible, and considerable areas where crops are grown but yields are low and variable. In Asia and North Africa irrigation has been a long established solution to this problem but irrigation and water control techniques played little part in the traditional farming systems of tropical Africa, with the exception of eastern Madagascar, although some riverine soils were used after seasonal flooding in Senegal, the Niger valley and Lake Chad. Irrigation has largely come as a result of external forces. Madagascar was settled by Indonesian migrants some 2000 years ago, and most other irrigation schemes were either a result of colonial government initiatives or were undertaken by European settlers. The Gezira scheme in the Sudan was begun in 1925 during British rule, although the irrigated area has increased substantially since independence; in Senegal the French established rice growing in the Richard Toll scheme, and also began the irrigation of the inland delta of the Niger in Mali; in Zimbabwe some 30,000 hectares are irrigated in the Chiredzi-Hippo valley triangle, a European farmed area. Some 28 per cent of the value of Zimbabwe's agricultural output comes from irrigated land. But although the area irrigated in tropical Africa has increased more rapidly than the expansion of the arable area since 1950, it remains an unimportant part of African farming; indeed in 1982 only 5 million hectares, about 2 per cent of sub-Saharan Africa's arable land, was irrigated, and 70 per cent of this was in Sudan, Madagascar and Nigeria. The irrigated area produces about 10 per cent of the region's cereals, and 6–8 per cent of its roots and tubers. Unfortunately there is limited potential; at most 20 million hectares could be irrigated, and the areas with the least potential are in the region where it is most needed, the Sahel. Current schemes are not well managed; the cost of each unit of water delivered is three to four times that in India.[43]

IMPLEMENTS, MACHINES AND LABOUR

Most African farmers still use relatively simple implements to farm the land. Land is cleared with axes and machetes and the vegetation burnt. In the more extensive systems planting is undertaken with a digging

Table 8.2 Share of different power sources in total power input for crop production, 1982–4 (percentage of total man-day equivalents)

	Labour	Tractors	Animals
Sub-Saharan Africa	89	1	10
Asia[a]	68	4	28
Latin America	59	22	19
Near East and North Africa	69	14	17
All developing regions	71	6	23

[a] Excluding China.
Source: N. Alexandratos (ed.), *World agriculture: toward 2000, an FAO study*, London, 1988, p. 145

stick, in the more intensive with broad blade hoes with short handles. Weeding is done with hoes, which are also used to harvest root crops, and the sickle or hands are used for harvesting the cereals. The use of machinery is rare, and tractors or harvesting machines are uncommon outside European farmed areas or on some large-scale settlement schemes (figure 8.9). The plough was unknown or unadopted in Africa south of the Sahara before 1900, and there was no tradition of using draught animals for ploughs or for pulling carts. Although the numbers of both ox-drawn ploughs and tractors have increased in the last forty years, most of the farm work is still done by human muscle, and draught animals make a far smaller contribution to farm work than elsewhere in the developing world (table 8.2). The early European settlers introduced ox-drawn ploughs and in some regions in the 1920s and 1930s they were adopted by Africans, notably in Uganda, where it was associated with the spread of cotton after 1910, and in parts of northern Rhodesia and Kenya.[44] In the West African savannah both British and French colonial governments advocated not only the adoption of ox-drawn ploughs but also the integration of livestock and crops, but with limited success. Thus in the Kusai district of north-east Ghana bullocks and ploughs were introduced in the 1930s, but by 1960 only 12 per cent of the households had them. In Senegal the number of ploughs increased rapidly and in 1972 there were 21,000; but this was on only 1–2 per cent of all holdings. Animal draught power is thought to be used on only 15 per cent of the sown area of West Africa.[45]

The slow progress of the ox-drawn plough is not surprising, for its adoption requires fundamental changes in bush fallowing and makes the establishment of a natural fallow slow. This in turn requires the development of some alternative means of maintaining soil fertility if yields are not to decline. The oxen do not provide enough manure for even a very small holding. Oxen need fodder, particularly in the period before the beginning of the rainy season when ploughing takes place; in this dry period fodder is in short supply. In Kenya it needs 4 hectares of land to feed two oxen; few peasant holdings have such an area available or the credit to buy and feed the oxen throughout the year. In the humid tropical areas the prevalence of the tsetse fly and the lack of suitable fodder crops have hampered the spread of the ox.[46]

The advantage of the ox-drawn plough is threefold. First, it enables farmers to cultivate a larger area; thus in Mali families with only the hoe can crop no more than 3 hectares, those with the plough 5-10 hectares. Second, by removing more weeds it can increase crop yields. Third, it may reduce the amount of labour necessary. But these are not always advantages. In the drier parts of Africa there is little agricultural activity for much of the dry season. Much labour is then needed when the rains begin and the land must be cultivated and sown as rapidly as possible. Later there is a further labour peak when weeding becomes necessary. Increasing the area sown by using the plough may lead to labour shortages during the weeding period. Given the cost of acquiring and feeding oxen, and given the abundance of labour in some parts of Africa, the economic advantages of ox-drawn ploughs over hand labour are not clear cut.

The role of labour in African agricultural production is of course paramount, tractors and oxen providing no more than one-fifth of the power used on farms. It would seem that there are few labour shortages if the figures for agricultural population densities are to be believed (figure 8.1); furthermore, the agricultural population of Africa has increased rapidly since 1950, by 124 per cent compared with 68 per cent in Asia and 36 per cent in Latin America (table 4.6). However, this is not necessarily a guide to labour availability and it has been suggested that labour shortages have restrained the growth of food output. First, in many parts of Africa, and particularly in the east and south, large proportions of the male population are temporarily absent in the cities or in mining areas. Thus some 60-80 per cent of the labour in African agriculture is undertaken by women.[47] Second, surveys of activities in African rural areas suggest that Africans work about 1000 hours a year compared with 2000 hours a year in European agriculture and 3000 hours in Asian agriculture. But in African communities much time is spent on associated activities such as processing and marketing crops.[48] Thus the labour input is less than the agricultural densities might suggest. Third, there are, of course, major spatial variations in population density, so that in some of the sparsely populated areas labour is insufficient to allow increases in output. Fourth, and perhaps most important, in most of savannah Africa there are seasonal bottlenecks in work, so that lack of labour does preclude the expansion of the area in crops and the amount of weeding that can be done. This presents considerable problems when both food and cash crops are grown. Thus in Chad the sowings of sorghum and cotton occur at the same time.[49]

It has often been believed that mechanization can overcome the problems of both spatial and seasonal labour shortages and fundamentally increase African food output. Under both colonial and independent regimes, there have been a series of land settlement schemes, particularly in the savannah zones, where the state has tried to develop mechanization, and in particular the use of tractors. For a variety of reasons, agronomic, economic and social, few of these schemes have been successful; the cost of providing spare parts, petrol and maintenance has too often been prohibitive. Thus in the Office of the Niger settlement scheme in Mali ox-drawn ploughs have largely replaced the original tractors, and in the Tanganyika groundnut scheme in the late 1940s the cost of servicing machinery and the problems of maintaining soil fertility once the bush had been cleared led to the early abandonment of the scheme.[50]

The lack of power, the limited use of machinery and the simple implements used in Africa restrict output per head. In 1980 the productivity of labour in Africa was only one-twentieth of that in Europe and only one-quarter of that in Asia.[51]

THE GROWTH OF FOOD OUTPUT: LIVESTOCK

Livestock products are a very small part of African food consumption; calories derived from animal foods make up only 7 per cent of all calories compared with 9 per cent in Asia, 17 per cent in Latin America and 33 per cent in Europe. Livestock products are a small proportion of the value of total agricultural output in most African countries, although there are some obvious exceptions; a large proportion of the population of Mali, Mauritania, Botswana and Somalia are nomadic pastoralists. In the last country some two-thirds of total agricultural output comes from animals. Compared with Asia or Latin America, cattle densities are low (table 8.3). This is partly explained by the distribution of livestock. The presence of the tsetse fly excludes cattle – except the indigenous dwarf cattle – from much of the humid tropical region and parts of the savannah areas. Thus half the total area of Tanzania is infested with tsetse, one-third of Uganda and one-quarter of Kenya. In the whole of Africa some 10 million square kilometres are infested, and it has been estimated that controlling trypanosomiasis

Table 8.3 Livestock in the developing regions

	Numbers per hectare agricultural land		
	1961–5	1987	Percentage change from 1961–5 to 1987
Cattle[a]			
Asia	0.44	0.34	−23
Africa	0.13	0.19	46
Latin America	0.31	0.41	32
Sheep and goats			
Asia	0.45	0.54	20
Africa	0.15	0.37	147
Latin America	0.11	0.20	82
Pigs			
Asia	0.24	0.36	50
Africa	0.005	0.01	100
Latin America	0.08	0.19	50
Total livestock units[b]			
Asia	0.7	0.58	14
Africa	0.17	0.23	35
Latin America	0.39	0.49	26

Livestock units are total numbers × 1 for cattle, × 0.1 for sheep and goats and × 0.5 for pigs.
[a] Including water buffalo.
[b] Including poultry.
Source: FAO, Production Yearbook 1976, vol. 30, Rome, 1977; Production Yearbook 1988, vol. 42, Rome, 1989

– the disease spread by the tsetse fly – would allow an increase of 140 million cattle.[52]

In contrast to Europe, or many parts of the developing world, the bulk of the livestock are kept not on farms where crops are grown, but by pastoralists who grow little or no crops and of whom many are nomadic. Nor are crops grown for their fodder; only 6 per cent of African food output is fed to animals. African cattle, sheep and goats are for the most part dependent upon the natural vegetation, and particularly savannah grasses, for their food. Fodder supplies are thus limited not just to the rainy season but to the earlier parts of that season, for much dry matter growth after the first flush of the season is indigestible.[53] During the dry season not only are the herds short of fodder, which they can partly overcome by migrating to distant areas or seeking pasture in seasonally flooded plains, but there is also a lack of drinking water. Not surprisingly cattle put on weight very slowly. For most African pastoralists the major product of cattle is milk, but in the absence of good pastures yields are very low. Thus among the Masai and the other pastoral peoples of eastern Kenya milk yields average only 150 litres a year compared with 3000 litres obtained on European-managed farms in the uplands.[54] It has often been argued that African pastoralists regard their livestock as a prestige symbol rather than a means of income. This may be so; they are certainly regarded, both by pastoralists and by sedentary farmers who own cattle, as a source of income to be sold only in times of extreme poverty. More cattle are sold for meat than might be thought, but the slaughter rate is still low – about 9 per cent in eastern Kenya.[55]

In spite of the low productivity of livestock in Africa, the numbers kept have increased substantially in the last thirty years, even though there have been substantial losses during drought years. Thus one-third of Botswana's cattle were lost in the droughts of 1961–6, and in the prolonged droughts of the early 1970s there were severe losses in the Sahel zone; in Mauritania the number of cattle fell from 2.5 million in 1968 to just over 1 million in 1973.[56] This long-run increase in livestock densities, greater in Africa than in Asia or Latin America (table 8.3), has led to widespread overgrazing. In most pastoral areas livestock are individually owned, but the land is held in common. Thus there is little or no incentive for individuals to keep their numbers at the carrying capacity of the range. In Kenya attempts have been made to overcome this by grouping pastoralists into co-operatives, where the cattle are collectively owned, or by marking the range out in individual ranches.[57]

Although the numbers of cattle have increased substantially since the early 1960s, the amount of meat and milk produced has increased more slowly, partly reflecting the limited market for meat outside the towns where incomes have grown more rapidly than in rural areas. Little progress has been made in raising productivity, either by promoting ranches with controlled grazing and supplementary feeding, or in integrating crops and livestock on the same farm; most of the progress has been achieved on European-owned farms in Kenya, Zimbabwe and Zambia.

THE CAUSES OF SLOW AGRICULTURAL GROWTH

It has been shown that although African food output has increased since 1950 it has been at a slower rate than in any other developing region. A majority of this increase has come from an expansion of the sown area; yields have increased less than in Asia or Latin America, and African yields are lower than in any part of the world. This, in turn, is due to the failure to adopt the inputs that have increased yields elsewhere. Over the last twenty years Africa has been subject to major famines, and in the late 1980s a greater proportion of the population have been estimated to be undernourished than elsewhere. Numerous explanations, which are reviewed here, have been offered to explain this.

Population growth

Until the 1960s world hunger was thought to be a result of the failure of food supply to keep up with population growth. Since then it has been accepted that for the world as a whole and in most regions food output has exceeded population growth, and an adequate supply of food is available in most countries. Undernutrition is now thought to be due to a lack of money to buy food. But this is not true in Africa. Since the early 1960s population growth *has* exceeded the growth of food output. Output in Africa has increased more slowly than elsewhere; in addition the growth of population has been higher than in any other major region (table 4.2). This is due to very high fertility, for although life expectation at birth has risen from under 40 in the 1950s to 50 at present, death rates remain higher than in any other region (table 4.3). But so too does fertility; marriage is universal and rates of remarriage high, whilst the average age of marriage varies, by region, from 17 to 22 years. Less than 5 per cent of the population is estimated to use contraception. In a recent survey in ten sub-Saharan countries the preferred size of family averaged 7.5, which suggests that fertility, which has not fallen since the 1950s, is unlikely to fall greatly in the immediate future.[58]

Population growth has had an adverse effect upon the environment in some parts of Africa. In many regions it has led to a reduction in the length of the natural fallow in bush fallowing systems, so that soil fertility is not restored, crop yields fall and the risk of soil erosion increases. This occurred in parts of Nigeria in the 1960s and 1970s. Elsewhere population growth has led to new settlement in areas where cultivation is risky: in Ethiopia and Kenya steep slopes have been cultivated, with subsequent erosion, and in Kenya and the Sahel crop growing has pushed into areas of very low and highly variable rainfall. Deforestation has been widespread, not only to create new agricultural land, but for timber and for fuel; 90 per cent of Africans use wood as fuel for cooking. Just over one-quarter of the continent is still forest covered, but it has been estimated that one-third of the original has been removed.[59]

Problems of land degradation have not been confined to the areas of crop production. There are some 50 million pastoralists in sub-Saharan Africa, and as their numbers have increased since 1950, so have their livestock. This alone could have led to overgrazing and a reduction of the carrying capacity of their land. But in some regions, such as the Sahel and Kenya, their grazing lands, which are common property and not held by individuals, have been invaded by settlers seeking land, and so the pastoralists have lost substantial areas of grazing with consequent land degradation.[60]

CLIMATIC CHANGE

In recent years it has become the practice to discount the importance of drought and other natural hazards in causing famine and to place the blame upon poverty, inadequate government responses, civil war or other social failures. Yet the famines in the Sahel in 1972–4 and the more widespread problems of 1983–5 were clearly triggered by drought, and food supplies did decline. The tragedy of the Sahel led to much study of the environmental and economic problems of that region; some writers argued that the desert was expanding, or that desert like conditions were being created on the margins of the desert by human misuse of the land; others went further and argued that the removal of the vegetation cover was permanently reducing rainfall. More recently some scepticism has been displayed about both these concepts, but in many parts of the African savannah the 1970s and 1980s have seen rainfall decline. In the Sahel, for example, rainfall has been consistently below the long-term mean since 1965 (figure 8.10). This has been exacerbated by the period of above average rainfall in the 1960s, when, encouraged by the wetter conditions, agricultural settlement moved northwards.[61]

AGRICULTURE AND ECONOMIC DEVELOPMENT

Few social scientists appear to believe that either population growth or climatic change have been of paramount importance in causing falling agricultural productivity or increasing hunger. Much more stress has been put upon government policies. In the immediate post-war period

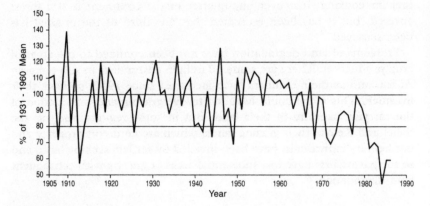

Figure 8.10   Mean annual rainfall for five stations in the Sahel expressed as a percentage of the 1931–60 mean *Source*: John T. Pierce, *The Food Resource*, London, 1990, p. 232

economists and governments in both Europe and the developing countries believed that the standard of living could only be raised by economic growth, and that this was most easily achieved by encouraging the development of manufacturing industry. It is not surprising then that after independence many African states tried to encourage industrial growth, although they all had an acute shortage of capital, skilled labour, managerial talent and often raw materials. Many countries produced national plans in which the improvement of industry and transport took priority, and the creation of health and educational facilities was also important. This left very little planned investment to be allocated to agriculture, although the majority of the population lived in rural areas and a high proportion of export earnings came from agriculture. It has been estimated that less than 10 per cent of capital expenditure in Africa is in agriculture compared with 20 per cent in India. Of individual countries, Tanzania allocated 12.5 per cent of government expenditure to agriculture in the 1970s, Senegal 2 per cent, the first national plan in Zambia 12 per cent, Sierra Leone 3–4 per cent, and even Kenya, with a comparatively successful agricultural performance, only 12.6 per cent. Malawi, Cameroon and the Ivory Coast were rare exceptions, the latter having allocated 28–30 per cent of public expenditure to agriculture, Cameroon 26 per cent, Malawi 19 per cent.[62] The food crisis of the early 1970s and the consequent steep rise in the price of food imports led some states to change their policies, in theory if not always in practice. Both Nigeria and Ghana have adopted plans aiming at self-sufficiency in food, and in Zambia the new national plan announced in 1980 allocated 30 per cent of the capital budget to agriculture.[63]

Export crops and government policy

In both Nigeria and Ghana the British government introduced marketing boards to control the prices and flow of the major export crops. The boards paid fixed prices to the peasant for cocoa or groundnuts and then sold the produce at the prevailing world price. It was hoped to supplement earnings in years of low world prices with the profits made in good years so that the peasant would be insulated from the fluctuation in world prices. First introduced in the 1930s, marketing boards dealt with individual export crops; the idea very rapidly spread to other British colonies and later was adopted in most French colonies. The marketing boards have been retained since independence, and indeed by the 1970s nearly all African export crops were controlled by marketing boards; attempts have also been made to extend the system to food crops.[64] The marketing boards have been subject to fierce criticism. It has been argued that, by retaining a high proportion of foreign earnings and paying peasants a much lower price, governments have heavily taxed peasants and reduced the incentive to increase output or improve productivity. In only a few countries – such as the Ivory Coast – have the marketing boards put earnings back into agriculture. The main beneficiaries have been not the farmers but a large white collar population living in the cities and employed by the marketing boards. Some have argued that the marketing boards have no useful function and should be abolished, allowing the price mechanism full play; this viewpoint was put forward forcibly in a World Bank report in 1981. During the 1980s the International Monetary Fund has required the development of free markets, and in some countries marketing boards have been abolished or private traders allowed to compete.[65]

Food crops and the government

In many African countries attempts have been made to control food prices in much the way that marketing boards have controlled export prices. In Africa since independence the major cities have seen what little industrial development there has been, and also a remarkable growth of administrators employed by the government. Political power has also been concentrated in the major cities, and governments have been eager to placate the urban populations. Thus attempts have been made to limit the increases in food prices, to the disadvantage of the farmers; urban wages have been subject to less restraint. There is thus a greater gap between urban and rural incomes than in other parts of the developing world. Whereas in Asia the ratio is 1 : 2 or 1 : 2.5, in Africa it is estimated to be between 1 : 4 and 1 : 9. African agriculture has had the 'role of a milch-cow . . . to support a dubiously productive urban elite in both public and private sectors'.[66]

Furthermore, import substitution policies that protect home industries and workers in industry have meant that the prices of farmers' purchases have risen far more than the prices farmers receive. Hence although African states have suffered from unfavourable terms of trade for their primary exports, especially since 1979, within the state the rural sector has suffered vis-à-vis the urban areas. This has greatly reduced the incentive for farmers to increase output or productivity. Indeed many economists believe that increases in food prices would lead to an increase in food production and resolve many of Africa's food problems; however, not all agree that raising food prices would increase the supply of food, for three-quarters of African food production is not sold off the farm, and without some improvement in agricultural technology substantial increases – whatever the price level – are impossible. In addition, farmers need to be able to buy consumer goods with the postulated greater incomes, and these are few.[67]

Research and extension

Much of the advance in output and productivity in the developed countries in the last forty years can be attributed to research in agronomy and the provision of extension services that have instructed farmers in the use of new methods, crops, fertilizers and pesticides. Such services are not widely available in Africa. Colonial governments established agricultural research stations in some African colonies early in the century; they were staffed almost entirely by Europeans and dealt mainly with the problems of export crops. Valuable work was done in the adoption of American cotton varieties in Uganda, on the cross-breeding of oil palm varieties, on the disease in the 1950s of cocoa plants and in Sierra Leone on new varieties of rice.[68] Interest in indigenous farming was limited until the 1930s when problems of soil erosion then attracted attention in the British colonies. In East Africa attempts were made to introduce conservation methods such as terracing, although with limited success. A more fundamental effort at change was made in northern Nigeria and north-east Ghana where attempts were made to promote the use of ox-drawn ploughs, the integration of livestock and crop production and the use of manures and rotations.[69]

Since the 1950s more attention has been paid to indigenous food crops, and research in this field has had better results than is sometimes stated. Hybrid maize suitable for the conditions of eastern and southern Africa have been bred and, in Kenya, widely adopted, and new varieties for the more humid conditions of southern Nigeria have been successfully tested.[70] But research in agriculture in Africa has been limited by lack of funds and by an acute shortage of African research workers. There is a further dilemma. The very high cost of imported technologies using fertilizers, pesticides and high yielding crop varieties suggests that research should be aimed at improving the indigenous farming systems, for they are adapted to the local ecological conditions. The need for replacing natural fallows with suitable rapid growing legumes is here paramount. But even if such traditional techniques were developed it is perhaps doubtful if they could provide yield increases large enough to keep up with likely future population growth. It is this which leads many to argue that the new technologies must be adopted in Africa if undernutrition and malnutrition are to be overcome.

The adoption of new technologies in Europe and America has been greatly assisted by the provision of extension services, and in particular of agents who can advise farmers about how to improve their methods. Attempts to provide similar services in Africa have foundered upon the lack of trained extension workers, who are often said to be of poor quality and are few in numbers; there is only one extension worker to every 3000 farmers in the West African savanna. With a shortage of extension workers to reach a large number of very small holdings, there has been considerable debate about the strategy of extension. Focusing on the bigger and better farmers in the hope that others will imitate their successes has not always worked and has often created a gap between the progressive farmers and the rest. Most of the extension work has been addressed to the male heads of households; yet women do a majority of the work on farms and indeed have been estimated to produce 85 per cent of Africa's food output.[71] The problems of reaching a large population of smallholders has doubtless prompted many African countries to create state-directed farms where, in theory, it is easier to introduce farming improvements. This has often diverted funds away from the small farmers who produce the bulk of the food. Thus in Tanzania in the early 1970s an attempt was made to relocate the dispersed rural population in villages on the grounds that it was cheaper to provide health and other welfare services to nucleated settlements; at the same time communal farming was encouraged for ideological reasons. It was felt that this would allow all farmers to be reached rather than a minority. But at the same time 80 per cent of all planned investment in agriculture went on large government projects, only 3 per cent to smallholders in villages.[72]

Table 8.4 Population size of tropical African states, 1989

Population classes (millions)	Number of states	Population (millions)
0–0.9	10	5.5
1–1.9	5	7.7
2–2.9	2	5.1
3–3.9	2	7.1
4–4.9	1	4.4
5–9.9	12	86.3
10–19.9	7	89.2
20–29.9	3	72.3
30.0 and over	2	158.1
Total	44	435.7

Government instability

African independence led to the creation of states based on the boundary lines drawn by European governments in the late nineteenth century. This has led to at least two severe disadvantages for the independent states. First, many of them have remarkably small populations. Tropical Africa's population of 435 million in 1989 was split into forty-four political units, of which twenty had individual populations of less than 5 million and a total population of only 30 million (table 8.4). This clearly provides many of these states with an inadequate market and duplicates the needs for research and agricultural services. Furthermore many of the boundaries cut tribal groupings and, perhaps more dangerous, contain feuding ethnic groups.

African states suffered from acute political instability; between 1960 and 1980 there were over twenty internal coups and fourteen wars. Not only did this lead to a quadrupling of real military spending – Chad and Ethiopia spend 10 per cent of their gross national product on defence – but it has prevented consistent economic policies being followed and has also led to physical destruction of rural areas.[73]

TENURE AND SOCIAL CONDITIONS

Thirty-five years ago most colonial authorities thought there were two primary obstacles to the increase of African food output: land tenure and the social attitudes of the population. Currently few writers put much stress upon these factors.

The principle feature of land tenure in traditional African life was the lack of individual freehold. Land was held communally by the tribe or village group, and land was allocated to individuals within the tribe by the tribal leader or village chief; those who cultivated a piece of land had the right to the products of that land, but not to sell it. Rent was not paid for the use of land and tenancy was unknown, except in Ethiopia where share cropping survived until 1974. Land that reverted to bush for fallow returned to communal ownership. Inheritance was not practised. There were doubtless greater variations in systems of land tenure than Europeans realized, but these essential features were widespread. This system had considerable advantages in a simple, sparsely populated society where there was little commercial production of crops and few things to buy. Above all it ensured that everyone had some land and that there were no great inequalities in the size of holdings, for land was allocated according to need.

But with European intervention in the twentieth century tenurial conditions became subject to stress. As population grew so the availability of land declined. This was compounded by the fact that in Kenya, Mozambique and southern and northern Rhodesia substantial areas were appropriated for the exclusive use of European settlers. The spread of cash crops, the use of money and the commercialization of agriculture also led to changes. But it was believed by most Europeans that the system of tenure inhibited improvements in agricultural productivity. The lack of freehold prevented farmers borrowing to improve their land; the fact that farmers did not farm the same plots of land over time inhibited permanent improvements to the land; and the allocation of land according to needs prevented the more efficient farmers increasing the area they could cultivate. In addition most African farms consisted of several scattered separate plots of land, and this reduced efficiency.

In Kenya the British colonial government consolidated holdings and introduced freehold in the former Kikuyu areas in the 1950s. Elsewhere overt changes in land tenure have been fewer, except where former European held lands have been subdivided among Africans, as in parts of Zambia, Kenya or Zimbabwe, or nationalized as in Mozambique; in the 1970s farmers in Tanzania were grouped into co-operatives, although there is now a move towards the adoption of long leases. In practice communal tenure has not inhibited progress. Thus the remarkable increase in the area in crops in the Ivory Coast was possible because it was allowed that the first person to clear new land had the right to its use. Elsewhere inheritance has become more common, particularly in the cash crop areas or near towns. Few authorities now believe that tenurial conditions are a major obstacle to progress.[74]

Although African farms remain predominantly small some

differentiation has emerged in the last forty years. Ironically, although this was once thought to be a measure of progress, it is now said to be a cause of social inequality and poverty. But differences in farm size are small compared with those in much of Asia and especially in Latin America. The number of large farms has probably increased, as has the use of hired labour; at the same time the other farms have been sub-divided as population has grown. But the progress of capitalist farming – if this is to be characterized by the emergence of large farms and a large landless labour force – has been relatively slow. It has been es-timated that smallholders produce 95 per cent of sub-Saharan food output; in Kenya, one of the few countries with a substantial large farm sector, 90 per cent of the chief foodstuffs are produced upon small holdings.[75]

While Africa remained under European rule many believed that the social characteristics of the African people, particularly in their tribal context, would hamper any improvement in agricultural output and productivity. Africans, it was argued, were not economic men in the way that European farmers were; profit maximization was not their primary aim and social obligations came before economic motives in determining behaviour; African farmers were thought not to respond to the incentives of price changes. Such views are no longer emphasized as obstacles to increases in output. It is not that African food producers do not respond to prices so much as that food prices have been held down by urban governments. Farmers do not ignore innovations for cultural reasons, but because they do not appear to give any substantial increase in output or are too costly to buy.[76]

DEPENDENCY AND EXPLOITATION

So far Africa's failure to raise food output sufficiently has been attrib-uted to the inadequacies of the traditional farming systems under rapid population growth and the failure to adopt new technologies, and these have been compounded by unsound economic policies. But there are those who believe that Africa's plight is a result primarily of colonial exploitation. Not only, it is argued, did Europeans seize African land in the south and east of Africa, but also, by introducing taxation and other means, they compelled Africans to work on European-owned estates and mines. Further, Africans after independence remained dependent upon the export of cash crops to Europe, for which there are no other markets. This policy has been supported in the post-independent period by local elites who benefit from connections, not with European governments, but with mining and agricultural marketing companies.[77]

The implication of such themes is not explored here, save in that they relate to food production. Cash crops do occupy a substantial proportion of African cultivated area (between 10 and 15 per cent), although not all of the output is exported. It has also been argued that they occupy a disproportionate amount of the better land. It would seem reasonable to suggest that the area devoted to export crops should be diverted to food production; this would enable food imports to be reduced as well as improving the inadequate consumption. But African states have a dilemma. Agricultural exports, in spite of their lack of buoyancy in the last decade, still provide a large proportion of foreign earnings, which are required for a wide range of goods other than food imports. A return to a policy of self-sufficiency may make matters worse.[78]

CONCLUSIONS

Africa has the least successful record of any of the developing regions in the post-war period. In 1950 available supplies of food were low – below, for the most part, estimated requirements – and malnutrition if not undernutrition was widespread. Since 1950 the population of tropical Africa has increased far more rapidly than any other region and there has been no sign of any reduction in the rate of increase. Food production in the immediate post-war period kept up with population growth, but since the 1960s the rate of increase has fallen and output per caput has fallen dramatically. Hence in 1986–8 only a handful of African countries had food supplies – including imports – that met needs even if supplies had been distributed according to requirements rather than by income or by possession of land (figure 2.4).

Most of the increased food output has come from increasing the area in crops, principally by reducing the fallow rather than by occupying uncultivated areas. Crop yields have increased little, and in some areas have fallen. Bush fallowing is still the prevailing agricultural system; this is a system that will maintain yields if the fallow is sufficient, but not one capable of dramatic increases in yield with traditional methods. In other parts of the world – and in a few parts of Africa – good increases in yields have been obtained through the use of chemical fertilizers, better water control and new crop varieties. Few of these inputs are used in Africa.

Behind these agronomic problems lie difficulties in government policies, a dependence upon export crops and a history of government instability, war and civil war. It is, perhaps, surprising not that Africa's food output has increased so slowly but that it has increased at all.

9

Latin America

Latin America's food problems are very different from those of Africa, yet there are a number of important similarities between the two continents. First, Latin America like sub-Saharan Africa is sparsely populated; both had a population density of 21 per square kilometre in 1989. Indeed the density of the agricultural population to the arable area (figure 8.2(d)) is noticeably lower than in either Africa or Asia. As in Africa population is concentrated in relatively few areas and there are large expanses with very low densities and hence an apparent abundance of agricultural land. The geographical distribution of population today reflects conditions found at the time of European arrival, when the Amerindian population was concentrated in the upland basins of Mexico, Central America and parts of the Andes. The Pacific coastlands were neglected except for the oases of the Peruvian desert, and there were few Indians in the rain forests of the Caribbean coast or the Amazon. The Spanish who conquered most of the continent in the sixteenth century needed labour as well as land and so settled initially in the densely populated areas. The Portuguese, who settled the sparsely populated coast of Brazil, brought slaves from Africa to work their sugar-cane plantations. Until the nineteenth century migration from Europe was comparatively small and mainly from Iberia; but then there was substantial immigration from both Iberia and other parts of Europe into Argentina, Uruguay and the southern states of Brazil.

In spite of the frontier movements of the last 200 years into the interior of Latin America, the present distribution still reflects that in 1500. The highest rural densities are found in the southern uplands of Mexico, western Guatemala, El Salvador and the volcanic meseta of Costa Rica; in the Andes there are major settlements in the intermontane uplands – the *altiplano* – from Colombia south to Bolivia. In Brazil rural settlement in this century has advanced away from the coast into São Paulo state, Parana and south towards Paraguay and Uruguay, but the major pattern remains unchanged. Half Latin America's population lies within 300 km of the coast; in few places are there densities comparable with those of rural Asia, except in parts of southern Mexico. The major change, apart from the Brazilian frontier, is the movement down from the uplands to the Pacific coast or to the Amazonian *selva*. Thus in Peru the sierra had 76 per cent of the national population in 1876, but this had fallen to 42 per cent in 1972, by which time the coast had more than half the population. In Ecuador the coast has also overtaken the sierra; it had 12 per cent of the population in 1856 and 54 per cent in 1972. Bolivia has no coastal region, and so it is the land to the east of the Andes that has gained at the expense of the *altiplano*; the *oriente* had 29 per cent of the population in 1976, compared with 20 per cent in 1950.[1]

The low population densities throughout much of Latin America have meant that, as in Africa, many farming systems have periods of fallow in the cropping sequence. Continuous cropping is unusual and multiple cropping rare except in the irrigated areas. However, the area in fallow has steadily declined; in the peasant farming areas population growth has reduced the land available and fallows have had to be reduced, whereas in the more advanced farming regions the use of fertilizers and irrigation has allowed continuous cropping. In the early 1960s 54 per cent of Latin America's arable land was in fallow, in the 1970s only one-third. In Mexico half the country's arable land was in fallow in the 1930s, but by 1960 this had fallen to 35 per cent; however, even in the densely populated rural areas of south central Mexico, 25 per cent of the arable was still in fallow. In such areas a year or two in crops is followed by one or two years in fallow, without any restoration of the natural vegetation; but in the more sparsely populated rain forest areas of Amazonia where isolated Indian communities survive, patches of land are cleared by slash and burn only once in twenty years or more. Such is the abundance of land – or more accurately the belief that land is abundant – that, as the Brazilian frontier penetrated into São Paulo and Parana in the first half of this century, land was sown to coffee and other crops and abandoned when yields declined. Farmers then moved onwards to reclaim new land, leaving eroded and thinly settled land behind them which generally became pasture. Only recently has this practice, which gave rise to the 'hollow frontier', given way to more intensive practices.[2]

A third characteristic that Latin America shares with Africa, and indeed with all the developing regions, is a dependence upon agricultural exports (figure 11.5). In 1987 agricultural exports formed 28 per cent of all Latin America's exports by value – a decline from 53 per cent in 1950 – and three-quarters of all these went to Western Europe and the United States. In only four countries were agricultural exports less than 25 per cent of all exports (table 9.1); in Bolivia and Venezuela tin and petroleum made up the majority. In Nicaragua, Paraguay and Cuba, in contrast, over 80 per cent of all exports were agricultural. Even in Brazil, which has a well-developed industrial sector and important exports of manufactured goods, agricultural exports were 28 per cent of that total.[3]

ECONOMIC DEVELOPMENT AND URBANIZATION

The contrasts between Latin America and Africa, however, are more important than the similarities, for Latin America has a higher level of economic development than Africa or most of Asia, and a very different farm structure.

The gross national product (GNP) per caput of most Latin American states was higher in 1950 than those in Africa or Asia, and remains so today (figure 4.1). Only Haiti falls into the low income bracket as defined by the World Bank. The richer countries – Mexico, Brazil, Uruguay, Argentina and Venezuela – have many features in common with the developed countries; indeed, thirty years ago Argentina and Uruguay were so classified in United Nations publications. Venezuela owes its high income per caput primarily to petroleum exports, but Mexico and Brazil have developed a substantial industrial sector in the last thirty years. In all these countries agriculture contributes less than 15 per cent of the gross domestic product (GDP), and in Chile, Argentina, Brazil and Venezuela the agricultural population is one-fifth or less of the labour force.

Latin America was more developed than Asia or Africa in 1950, or indeed in 1900. In the post-war period the region experienced a moderate rate of economic growth, but in the late 1970s this began to falter. Latin America had acquired a massive debt in the 1970s, and by the early 1980s was having difficulties in paying interest. In 1987 the region's total debt was $410.5 billion and paying interest on this took 30 per cent of all exports. Although the GNP grew in the 1980s it was at a very sluggish rate, and GDP per capita fell on average at 2 per cent per annum between 1980 and 1987.[4]

In Argentina, Uruguay and Cuba crude birth rates have fallen towards European levels, although in most of Latin America fertility remains high and the rate of population growth is nearer to that of Africa than of Europe (table 9.1). Levels of urbanization, however, are more like those of Europe. Much of the early Spanish settlement was urban, and a much higher proportion of the population was living in towns in 1950 than in Africa and Asia (table 4.5). This contrast persists:

Table 9.1 Economic and social characteristics of Latin American countries, 1988

Country	GNP per caput (US dollars)	Percentage of GNP from agriculture	Agricultural exports as a percentage of all exports	Annual growth of population, 1980–8	Crude birth rate per thousand	Percentage of labour force in agriculture	Urban population as a percentage of total
Haiti	380	31	28.8	1.8	35	65	29
Honduras	860	25	67.8	3.6	39	56	42
Bolivia	570	24	9.9	2.7	42	43	50
Nicaragua	n.a.	21	92.5	n.a.	41	40	59
El Salvador	940	14	64.4	1.3	36	38	44
Peru	1300	12	9.7	2.2	31	37	69
Dominican Republic	720	23	43.9	2.4	31	38	59
Colombia	1180	19	47.3	2.1	26	29	69
Guatemala	900	n.a.	68.1	2.9	40	52	33
Ecuador	1120	15	28.1	2.7	32	32	55
Paraguay	1180	30	91.0	3.2	35	47	46
Panama	2120	9	38.9	2.2	26	26	54
Cuba	n.a.	n.a.	85.3	n.a.	n.a.	20	n.a.
Mexico	1760	9	11.7	2.2	28	31	71
Chile	1510	n.a.	12.9	1.7	23	13	85
Brazil	2160	9	28.1	2.2	28	26	75
Costa Rica	1690	18	57.5	2.3	27	25	45
Uruguay	2470	11	42.4	0.6	17	14	85
Argentina	2520	13	61.7	1.4	21	11	86
Venezuela	3250	6	0.3	2.8	30	12	83

n.a., not available.
Source: World Bank, *World Development Report 1990*, Washington, D.C., 1990

in 1985 69 per cent of Latin America's population lived in urban places compared with 29 per cent in Africa and 28 per cent in Asia (table 4.5). The level of urbanization in Argentina and Uruguay is comparable with Western Europe, that in Cuba, Chile and Venezuela not far short. In 1988 only Haiti, Honduras, El Salvador, Guatemala, Paraguay and Costa Rica had a majority of their population living in rural areas (table 9.1). The rapid growth of the urban population since 1950 – it has risen fourfold – is partly due to high fertility and falling mortality rates in the cities, but this has been combined with much migration from the countryside, especially into the very large towns. It has been estimated that some 40 million people left the country between 1950 and 1975, attracted by the higher wages and better public services. In the country not only are employment opportunities fewer, incomes lower and public services such as education, public health and welfare poorer, but in some regions the landlord class can still exert great and repressive power over the peasantry.[5]

Because so many of the rural immigrants are young, and have married and had children in the towns, the towns have gained by their natural increase, and the rural areas have lost a substantial part of their population in the child bearing ages. Consequently by the 1970s the urban population of Latin America was increasing by 7 million a year, the rural population by only 1.5 million, and so the rate of increase of the rural population has been less than that in Asia or Africa. Indeed in parts of Latin America the rural and agricultural population has stagnated or even declined. In most Latin American countries the national rural populations have increased since 1950; but when populations are studied at the local level rural depopulation can be seen to be occurring in – among others – Peru, Ecuador and Colombia. In the last country the urban population rose fivefold between 1938 and 1973, and from 31 per cent of the total population to 61 per cent. The rural population increased absolutely in the country as a whole, from 7 million in 1951 to nearly 9 million in 1973. But numerous administrative districts in the older, densely settled areas suffered absolute declines in their populations in the 1960s and 1970s. This was primarily due to migration to the towns, but there was also movement to settle in the foothills of the Andes. This movement away from the older upland rural areas to either the *selva* east of the Andes or to the Pacific coastlands is found from Colombia south to Bolivia and also in Central America. Combined with the flow to the towns it has led to local rural depopulation in parts of Latin America, a phenomenon not yet found in Africa or Asia.[6]

Most of the rural population are dependent upon agriculture for their livelihood, and the trends in the rural population have effected changes

Table 9.2 Paid employees as a percentage of the agricultural labour force, Latin America

	Latin America	Brazil	Mexico	Colombia	Other
1950	35.3	33.7	30.3	43.5	38.3
1960	33.6–35.8	25.3–30.3	43.7	42.2	37.1
1970	34.2–35.9	25.4–29.2	48.2	46.4	35.5
1980	–	33.9	–	43.4	–

Source: A. Berry, 'Agrarian structure, rural labour markets and trends in rural incomes in Latin America', in V. L. Urquidi and S. T. Reyes (eds), *Human Resources, Employment and Development*, vol. 4, *Latin America*, London, 1983, p.180

in the agricultural population, which has increased by only 36 per cent between 1950 and 1989, far less rapidly than in Asia or Africa (table 4.6). This population can be divided into those who have no land and rely upon wage labour and those who have some land, however small. The number of small holdings has increased because of the subdivision of existing holdings as a result of rapid population growth, and also because of their creation in areas of colonization. On the other hand there has been a decline in the number of small tenants and share-croppers; in the 1960s many haciendas were converted to modern commercial farms run as one unit rather than split into tenanted farms. Since the 1960s this modernized sector has replaced men with machines and permanent workers by temporary workers; in Brazil, Mexico and Colombia only 10 per cent of the agricultural labour force are permanent employees. There were thus divergent trends in labour utilization between 1950 and 1975. In Brazil the amount of labour used per hectare increased on small holdings, remained constant on medium-sized farms and declined by one-third on those holdings over one hundred hectares. Somewhat surprisingly – probably due to the much greater size of the subsistence or peasant sector – there has been no great increase in the relative importance of the paid worker in Latin America since 1950; paid employees made up 36 per cent of the agricultural labour force in both 1950 and 1970, whilst the proportions in Brazil and Colombia showed little change between 1950 and 1980[7] (table 9.2).

Since the 1960s two distinct sectors have emerged in the economy, one consisting predominantly of small holdings with a large subsistence element and mainly traditional farming methods, and the other modernized, using new inputs and operating on a large scale.

These two sectors have quite different problems. In the peasant sector, incomes are low, farms are small and subdivided, steep slopes are cultivated with subsequent soil erosion and families have to seek employment off the farm, often by migratory labour. Thus in the uplands of Guatemala farmers get about half their total income by working on the large farms of the Caribbean and Pacific lowlands during harvest time, and even since the land reforms of the 1970s Indian farmers from the Peruvian sierra still move to the oases of the coast to work on the sugar and cotton plantations – now co-operatives. The peasant sector has absorbed much of the rural population growth since the 1950s; the subdivision of the farms has meant that a considerable proportion of the agricultural workforce is underemployed, in 1970 between one-fifth and one-third. The smaller number of landless labourers, perhaps one-third of the total agricultural workforce who work in the modernized sector, have had different problems. As land reform measures have ended the servile conditions that existed in the 1950s, when *colonos, peons* and others received small plots of land in return for working on hacienda, so many landlords ended the provision of land and substituted wages; and as minimum wage legislation has been slowly introduced, so landlords have replaced labour with machines. One noticeable trend has been for the permanent labour force to be pared to the minimum, temporary labour being hired in the periods of heavy labour needs. So rapid has been the growth of the urban population in the last forty years that there are few opportunities for employment left in the cities for the impoverished rural populations. Not surprisingly it has been calculated that in 1980, 56 per cent of the rural population of Latin America lived below the poverty line compared with 23 per cent of the urban population, and that in 1970 average urban incomes were three to four times those in the rural areas.[8]

LAND DISTRIBUTION AND LAND REFORM

The industrialization of parts of Latin America and the large-scale flight to the cities has meant that the increase of the agricultural population has not been at such a rapid rate as in Africa or involved such large numbers as in Asia, although of course it has given rise to grave difficulties. But underlying many of the agricultural problems of Latin America today is the distribution of land and the farm size structure, which is very different from any other part of the world.

In the sixteenth century the Spanish and Portuguese conquerors allocated very large grants of land to individuals. Indians who lived on these holdings were required to work the land and often to undertake other services such as domestic work; in return they were allowed to use small plots of land. In some areas Indian communities at first retained much of their land, but this was often expropriated in the nineteenth century. In Brazil the Portuguese imported slaves from Africa to work the sugar plantations, and slaves were also taken to other parts of Latin America. The hacienda, where very large units of land were cultivated by *peons, colonos, inquilinos* or *huasipungos* in return for land in small amounts for their subsistence, became the typical Latin American system of land ownership and operation, although there were many variations. Thus in the Peruvian *altiplano* nearly all of a landlord's estate would be divided among the Indian population, and his income came in the form of rent. More commonly *peons* worked the land under the supervision of the landlord or a manager, while supporting themselves on small plots of land either on or near the hacienda. After the end of slavery in Brazil share cropping became a typical form of land holding in the north east, and in the south coffee *fazendas* were also often worked by share croppers. What was conspicuously absent was the medium-sized family farm that characterized much of North America or Western Europe. There were some exceptions. Medium-sized owner-occupied farms were established in the upland area of Costa Rica, although they have diminished in importance in this century, and the colonization of Antioquia in Columbia was partly by such farmers. In Haiti slavery and the plantation gave way to small owner-occupied farms after the revolution of 1798. In Argentina and more so in southern Brazil many of the European immigrants of the late nineteenth century eventually gained title to small and medium-sized farms, and there are many family farms in the states of Rio Grande do Sul and Santa Catarina. But in most of Latin America there remains a great difference between the very large number of small farms occupying a small proportion of the farmland, and the very few holdings occupying a large proportion of the area.[9]

Statistics on the ownership of land, the size of farms and the number of landless labourers are notoriously unreliable in Latin America, and comparisons between countries are difficult to undertake. There is often a confusion between the size of units of ownership and that of units of operation, whilst in some countries those with small plots of land received for working on the hacienda are recorded as labourers, in other as farmers.[10]

The distribution of land in a few countries in 1970 demonstrates the predominance of the small holding as a proportion of all holdings, and of the large holding as a percentage of the cultivated area (table 9.3). The dominance of the large holding was particularly marked in Argentina and Venezuela. Figures for the whole of Latin America, dating

Table 9.3 Size of holdings in selected countries, Latin America, c.1970

	Percentage of holdings					Percentage of area				
	0–5 hectares	5–20 hectares	20–100 hectares	100–1000 hectares	1000 + hectares	0–5 hectares	5–20 hectares	20–100 hectares	100–1000 hectares	1000 + hectares
Argentina	18.2	23.0	26.0	26.6	6.2	0.1	0.8	3.9	20.6	74.6
Brazil	36.6	30.2	23.6	8.5	0.7	1.3	5.5	16.7	36.9	39.6
Colombia	59.5	23.6	12.6	4.0	0.3	3.7	8.6	20.2	37.0	30.5
Venezuela	42.9	31.9	16.4	7.1	1.7	1.0	3.1	6.9	22.0	67.0
Mexico	61.0	20.4[a]	11.0[b]	6.6	1.0	1.2	3.5[a]	8.5[b]	27.1	59.7
Honduras	64.0	25.0	10.0	2.0	–	9.0	18.0	30.0	29.0	22.0
Ecuador	69.7	18.5	9.9	1.9		7.8	13.0	28.7	50.7	

[a] 5–25 hectares.
[b] 25–100 hectares.
Source: Alain de Janvry, The Agrarian Question and Reformism in Latin America, Baltimore, Md., 1981, pp. 96–7.

Table 9.4   Farm structure in Latin America, c.1960

Farm size	Number of holdings		Area	
	thousands	per cent	million hectares	per cent
Under 20 hectares	5445	72.6	27.0	3.7
20–100 hectares	1350	18.0	60.6	8.4
100–1000 hectares	600	8.0	166.0	22.9
1000 hectares and over	105	1.4	470.0	65.0
Total	7500	100.0	723.6	100.0

Source: J. Chonochol, 'Land tenure and development in Latin America', in C. Veliz (ed.), *Obstacles to Change in Latin America*, Oxford, 1965, pp. 79–90

Table 9.5   Small and large holdings in the developing countries, 1970

	Small holdings			Large holdings		
	Numbers (%)	Area (%)	Average size (ha)	Numbers (%)	Area (%)	Average size (ha)
Latin America[a]	66.0	3.7	2.7	7.9	80.3	514
Africa[b]	66.0	22.4	1.0	3.6	34.0	28
Near East[c]	50.0	11.2	1.6	10.3	54.7	50
Far East[b]	71.1	21.7	0.7	4.0	31.1	17

[a] Small, below 10 hectares; large, above 100 hectares.
[b] Small, below 2 hectares; large, above 10 hectares.
[c] Small, below 5 hectares; large, above 20 hectares.
Source: P. Harrison, 'The inequities that curb potential', *Ceres*, **81**, 1981, pp. 22–6

however from 1960, confirm this pattern of land holding. In 1960 two-thirds of the agricultural land in Latin America was occupied by holdings of 100 hectares or more, which made up only 1.4 per cent of the number of holdings (table 9.4). Conversely three-quarters of all holdings were less than 20 hectares, but occupied only 3.7 per cent of the agricultural area. Such inequality is not to be found elsewhere. Thus in Africa in 1970 (table 9.5) small holdings of less than 2 hectares made up two-thirds of all farms, but they occupied one-fifth of the farmland; in the Far East small farms were 71 per cent of all holdings but also occupied one-fifth of the farmland. Large holdings nowhere – in the developing world – occupy such a large proportion of the total farmland as in Latin America, and their average size in Africa and Asia is much smaller (table 9.5).

Because of the inconsistencies in the way that workers with small plots of land are recorded, it is difficult to establish what proportion of the Latin American rural population is without land; it has been put variously as 25, 33, 40 and 63 per cent. However, a comparison of landlessness made by the Food and Agriculture Organization (FAO) in the 1970s suggests that 34 per cent of Latin America's agricultural population was then without land, compared with 31 per cent in the Far East, 25 per cent in the Near East and 10 per cent in Africa.[11]

In the past the landowners of Latin America had great financial and political power, and the prospects of altering the distribution of land were remote. In this century there have been many attempts to redistribute land, some of which have been successful. The Mexican revolution was followed by legislation in 1917 that gave the rural population entitlement to claim land, and also specified the sort of land that would be expropriated. It was not until the 1930s that expropriation and allocation took place, but by 1980 some 77 million hectares had been distributed. Most of this was given not to individuals but to communities called *ejidos*. Although land was generally worked by individuals – only a few *ejidos* were collectively operated – *ejido* land could not be sold or rented out. About one-half of the agricultural area of Mexico is *ejido* land and one-half privately owned. In Bolivia the revolution of 1952 was followed by peasant invasions of hacienda land, and the legislation of 1953 followed rather than caused expropriation. The Cuban revolution ended private ownership of the sugar plantations but did not lead to land redistribution, although the status of the labourers on the state farms was improved. In the 1970s the military government of Peru nationalized the foreign-owned plantations of the coast and expropriated hacienda land in the sierra; nearly 8 million hectares had been allocated to peasants by 1976, and only 16 per cent of the Peruvian agricultural population remained landless. More recently land has been expropriated in El Salvador and Nicaragua; in Chile the reforms carried out under Allende have been reversed. Although there has been little significant land redistribution elsewhere, the servile status of tenancies – such as that of the *huasipungo* in Ecuador – has nearly everywhere been abolished. In most countries – particularly those where land redistribution has not taken place – governments have encouraged the colonization of new land, perhaps as an easy alternative to the problems of expropriation.[12] Although land reform has reduced the importance of large farms in a few countries, they remain the dominant unit of production in Latin America, but are dependent now upon wage labour rather than *colonos*. The principal change in farm size has been the great increase in the number of small farms and the reduction of their average size. Thus, for example, in the sierra of Peru the number of holdings less than 5 hectares rose from 500,000 to 884,000 between 1961 and 1972, and in Guatemala farms of less than 0.7 hectares were only one-fifth of all holdings in 1950 but 41 per cent by 1979. In Brazil the average size of those holdings less than 5 hectares fell from 4.25 hectares in 1960 to 3.6 hectares in 1970, and in Ecuador from 1.7 hectares to 1.5 hectares. This increase in the number of small farms has had various causes. Existing holdings have been subdivided by population growth, and many of the farms allocated in reform programmes have been small; similarly much of the settlement in pioneer zones has been on a small scale. Indeed in many areas of colonization the latifundia–minifundia pattern has reappeared.[13]

LAND USE IN LATIN AMERICA

It is over 5000 miles to Cape Horn from the United States border with Mexico. Not surprisingly Latin America includes a diversity of environments and types of farming, although neither are as well described as might be wished. The core of Latin America is the Amazon basin, where high temperatures throughout the year and heavy rainfall – with only locally a significant dry season – sustains the *selva*, or rain forest, the most luxuriant vegetation type on earth. In spite of recent attempts to exploit the rain forest, it remains sparsely populated. Rain forest or tropical forest also occurs on the Caribbean coast, south of Vera Cruz, and the original vegetation of the Brazilian coast from Recife to Rio de Janeiro was tropical forest, although far less luxuriant than the rain forest and with a significant dry season. Between the Amazon and the Brazilian coast lie crystalline plateaux with a marked dry season and a vegetation adapted to seasonal drought, a combination of grass and trees; in the north-east interior of Brazil the *caatinga* is drought resistant. Southwards towards Paraguay grass is the dominant natural vegetation, as it was in much of Argentina and Uruguay where a subtropical climate and good soils sustain the most productive farming systems of the continent. Southwards into Patagonia aridity precludes crop production, and in the extreme south, in Argentina and Chile, are found the only lowland cool temperate climates of Latin America.[14]

The most distinctive feature of the continent is the upland belt which stretches from Mexico to Chile. This region has two important characteristics. The numerous volcanoes provide fertile soils, and the altitude reduces temperatures and thus gives rise to distinctive zones where crops of very different climatic regions are grown quite close to each other. For the most part, however, the upland areas suitable for agriculture, although locally giving rise to high rural densities, are small in extent and separated by terrain difficult to traverse. In some parts of the Andes settlement occurs at heights where temperatures are so low as to only allow the growth of the more hardy temperate crops such as the potato or barley, and south from northern Peru the intermontane basins suffer not only from frost but also from drought, particularly in southern Bolivia, where the only possible form of agriculture is sheep rearing.

The Pacific coastlands of Latin America, generally very narrow, although for the most part hot, all experience a marked dry season. From Ecuador southwards this becomes a desert where agriculture is only possible using the water of streams descending from the Andes, most noticeably in northern Peru. Further south, central Chile has a climate and agriculture comparable with that of the Mediterranean basin.

Classifications of Latin American agriculture are apt to be based upon its farm structure; there are few typologies based on land use, and even fewer attempts to map the distribution of farm types. Aggregate figures for Latin America can be deceptive for, although there are thirty-nine independent states (if the Caribbean countries are included), Brazil, Argentina and Mexico account for two-thirds of the total land area, three-quarters of the arable land and 60 per cent of the population. Some of the essential features of land use are shown in table 9.6. Although much emphasis is usually placed upon the importance of export crops, it is the food crops that occupy most of the cropped area. Maize is widely grown and is the most important crop. Wheat, although the second cereal crop, is largely grown in Argentina and southern Brazil. In the latter region wheat is increasingly grown in rotation with soybean. This crop has expanded remarkably in the last thirty years and Brazil is now the second most important world source of soybeans and soybean products. Rice has also expanded considerably. It is widely grown in Latin America, not as an irrigated crop – only 30 per cent of the total rice area is irrigated – but as an upland, dry crop; the bulk is grown in central and southern Brazil.

Root crops, although occupying a comparatively small area, are regionally significant; potatoes are important in the Andean *altiplano* and manioc (cassava) in many lowland shifting cultivation regions. Fruit and vegetables occupy a considerable area, but particularly in the environs of the major cities. The traditional export crops of cocoa, sugarcane and coffee (and bananas and cotton, for which output but not area data are available) occupy rather less of the total arable than might be expected, although locally – particularly in Central America – they are often dominant in the land use pattern.

Table 9.6 Land use in Latin America, 1961–5 and 1989 (million hectares)

	1961–5	1989
Arable area	116	180
Pasture	493	569
Forest	1061	960
Irrigated land	10	15
All cereals	40	52
Wheat	8	10
Rice	5	8
Barley	1	1
Maize	22	28
Sorghum	2	4
All root crops	3.5	4
Potatoes	1	1
Sweet potatoes	0.4	0.3
Cassava	2	2.5
Pulses	6.3	10.0
Soybeans	0.4	16.0
Vegetables	9.6	n.d.
Sugar-cane	4.5	8.0
Bananas	n.d.	n.d.
Coffee	4.3	6.0
Cotton	n.d.	n.d.
Cocoa	1	1.5

n.d., no data.
Sources: FAO, Production Yearbook 1989, vol. 43, Rome, 1990; Production Yearbook 1976, vol. 30, Rome, 1977

Latin America has more livestock per caput of the total population than any of the major regions, although the density is low; pasture occupies three times the area in crops and the tradition of extensive livestock raising is long established. The Spanish and Portuguese brought long-horned cattle to the Americas in the sixteenth century, and also Iberian methods of livestock raising – the open range, the annual round-up and branding and the mounted herdsmen. Ranching became a major form of land use in northern Mexico in the seventeenth and eighteenth centuries, as it did in the *cerrado* of Brazil, whereas in the late nine-teenth century a more intensive system of livestock production devel-oped in Argentina and Uruguay. In much of Latin America the quality of the cattle is low, European breeds having been adopted only in Argentina and Uruguay and on a few ranches in northern Mexico. Not only are densities low, but feeds other than the natural vegetation are uncommon, cattle take six years to reach slaughter weight and losses from disease and drought are high. Dairying is a recent development outside the temperate south, and is mainly found near the cities, particularly in southern Brazil. Indeed in São Paulo state the value of livestock products now exceeds that of coffee.[15]

There have been few attempts to establish maps of types of farming in Latin America; indeed there are few systematic accounts of the agricultural geography of the continent. However, the most useful typological distinction is probably between the Indian and *mestizo* small-scale farms, producing mainly food crops, that are to be found in the plateaux of Mexico, Central America and the Andes, and the large-scale hacienda and plantations, often producing one crop only and oriented to export. But in southern Brazil and parts of Argentina European immigrants have established farms producing a variety of products – some for home consumption and some for export – on a much smaller scale than the traditional hacienda, but efficiently and with often highly mechanized methods.[16]

THE GROWTH OF FOOD OUTPUT

In the 1950s Latin American agriculture was widely described as backward, and it is not without its critics today. Yet food output has increased rapidly since the end of the Second World War, at 3.1 per cent per annum in the 1950s, 3.5 per cent in the 1960s and 3.8 per cent in the 1970s, although falling to only 2.2 per cent in the 1980s. But Latin America, like Africa, has had a very high rate of population increase: 2.8 per cent per annum in the 1960s, 2.7 per cent in the 1970s and 2.2 per cent in the 1980s. Consequently the rate of food output growth per caput has been less impressive. Indeed in a number of countries, including El Salvador and Ecuador, it declined between 1961–5 and 1976, but these countries held a population of 52 million in 1980, only 14 per cent of the population of Latin America.

In the 1980s declines in food production per capita have occurred in more countries, as the growth of food production has fallen well below the rate of growth in previous decades; this has been particularly so in Central America. Even so there are few countries in which national food supplies fall below minimum requirements (figure 2.4); the situation in Latin America is very different from that in Africa.

Although there have been increases in crop yields in many parts of Latin America since 1950, most of the increased food output has come from the expansion of the area in crops. The arable area more than doubled between 1950 and 1988, although the reservations on the definition of arable land should be recalled (see p. 75); furthermore some of this increase was in non-food crops. But the area in the major food crops recorded by the FAO indicates a doubling of the area between 1950 and 1988 (table 5.3), a figure confirmed by other estimates. Some two-thirds of the extra food produced in Latin America between 1950 and 1980 came from the expansion of area; the increase due to higher yields has, however, slowly increased with time. Thus between the mid 1930s and the mid 1950s crop yields accounted for only 20 per cent of the extra food, in the 1960s 30 per cent and in the 1970s 40 per cent. The way in which the area has been expanded has been quite different from either Africa or Asia. Multiple cropping is rare in Latin America except in some irrigated areas; but only 9 per cent of the arable is irrigated and one-third of this is in Mexico. The area double cropped in Mexico rose from 40,000 hectares in 1950 to 800,000 hectares in 1980, but this was only 15 per cent of Mexico's irrigated land. Indeed a substantial part of this is left fallow each year owing to problems of salinity. Nor has the reduction of fallow been a major contributor to the increased area in crops as it has been in Africa, although it is true that the fallow has been much reduced in parts of southern Mexico, Central America and the Andean *altiplano*. Of most importance has been the colonization of new land, a matter touched upon in an earlier chapter but of so much importance in Latin America that it merits further consideration.[17]

THE COLONIZATION OF NEW LAND IN LATIN AMERICA

It will be recalled that the Indian and early Iberian settlement of Latin America was confined to the uplands of Central America and the Andes and the coasts of Brazil. The lowland rain forest of Central America and the Amazon basin had a sparse population, practising a combination of shifting agriculture and gathering and hunting, whereas the drier *cerrado* and pampa areas of Brazil, Uruguay and Argentina were occupied by extensive ranching, as was much of northern Mexico and the *llanos* of Venezuela. Three events had led to the integration of these regions into the economic life of Latin America and to their denser settlement. The first was the development of export commodities for Western Europe and the United States, notably bananas from the Caribbean coast of Central America, coffee from São Paulo and meat and later wheat from Argentina. These were not of course the first exports from Latin America, but in the late nineteenth century capital from Europe and the United States became increasingly important, and there was also considerable emigration from Europe to southern Brazil and Argentina. Second, a beginning was made early in this century in the conquest of the diseases endemic to the tropical lowlands, although it was not until after the Second World War that malaria was finally eradicated, removing an obstacle to permanent settlement. The third was the slow spread of road and rail, without which it was difficult to get people in or goods out. The most extensive railway building was in Argentina, but railways were built elsewhere; indeed the banana boom in the 1890s was a consequence of the railway built from the Caribbean coast to San José, the inland and upland capital of Costa Rica. In the last thirty years railways and, more important, highways have been built to link the Andean altiplano with the Amazon territories in Peru, Ecuador and Colombia, prompted by military needs and the exploitation of oil but also allowing agricultural settlement. In the 1970s the Brazilian government undertook an extensive programme of road building in the Amazon.

Development of the remoter regions was on the whole slow until the 1930s and 1940s, but since then new land has been added in marginal environments at a prodigious rate. In Mexico the government invested heavily in irrigation in the north west; whilst the tropical lowlands on the Caribbean coast, largely undeveloped before 1940, have been the scene of considerable settlement, both spontaneous and government backed, large-scale and small scale, partly subsistence, partly export oriented. Indeed the Gulf states now provide one-quarter of Mexican agricultural output. The arable area of Mexico increased by 50 per cent between 1932 and 1970, although there has been little increase since the mid 1970s. The Andean and Central American republics have seen settlement of their Pacific lowlands, prompted in both cases by the eradication of malaria and the building of the Pan-American highway; this has been particularly notable in Ecuador, where the Pacific lowland is more humid than that to the north and south. Settlement here was prompted initially by the banana boom in the 1940s and 1950s but other crops, including rice, have become important; between 1954 and 1974 0.75 million hectares were cleared for cultivation and the coast has replaced the sierra as the economic centre of the country. In Peru and Bolivia the movement has been eastwards for the Peruvian coast is arid with few prospects of extending the irrigated area and Bolivia has no Pacific coastland. The Bolivian *altiplano* was densely populated in the north and at the time of the 1953 revolution there was little development in either the eastern Andean hills – the *yungas* – or the lowland *selva* region. However, these areas have been rapidly developed by both small-scale Indian settlement and large-scale capitalist production, so that the province of Cochabamba now has one-third of Bolivia's cultivated land; between 1950 and 1973 80 per cent of the increase in agricultural production was accounted for by Santa Cruz and the *yungas*.[18]

The most dramatic frontier developments, however, have come in Brazil and Argentina; the remarkable expansion of beef and wheat production in Argentina paralleled the development of Canada and Australia earlier in this century, but in the 1950s this expansion halted. Not so in Brazil. From 1900 onwards the state of São Paulo was occupied by coffee production, the frontier moving rapidly westwards and into northern Parana in the 1930s. Occupation of the southern states of Santa Catarina and Rio Grande do Sul – partly by European immigrants, partly by Brazilians – began in the late nineteenth century, but the rapid growth of the arable area, including the cultivation of rice, wheat and soybeans, has been mainly since the 1950s. Nor has the expansion of the cultivated area been confined to the progressive south. The area of food crops in the north east, poverty stricken, drought ridden and apparently undynamic, doubled between 1950 and 1968. More dramatically the movement of the capital to Brasilia in the *cerrado* has prompted the movement of crop cultivation into the once exclusively livestock region of Goias, southern Mato Grosso and Minas Gerais. Lastly has come the much heralded frontier in the Amazon rain forest, as yet not very productive of crops. The addition to the Brazilian cropland has been formidable. In 1920 6.6 million hectares were cultivated, but in the 1970s this had risen to 30 million, and expansion has continued up to the present. However, it is perhaps easy to exaggerate the significance of this. In fact it is only in parts of the south that more than 40 per cent of the land is in crops; most of the country remains unexploited (figure 9.1).[19]

TECHNOLOGICAL CHANGE

In 1950 the technological level of farming in much of Latin America was low. Rotations were rare – although on peasant farms intercropping was practised (see p. 103) – and monoculture characterized many areas of export crops, the land being cultivated until exhausted and then the farmers moving on to new areas. Although the continent had a large population of cattle and sheep, livestock and crop production were rarely combined on the same farm, livestock manure was not an important source of plant nutrients and few chemical fertilizers were in use. Farm implements were for the most part simple. Only 15 per cent of Latin America's cropland was then worked with tractors; and

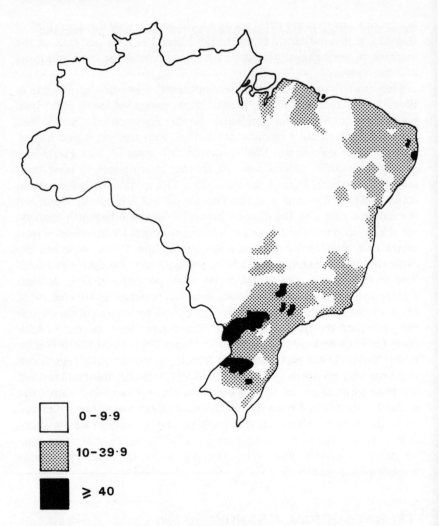

Figure 9.1 Percentage of total agricultural area cultivated, Brazil, 1970
Source: B. Bret, 'Donnes et réflexions sur l'agriculture Bresilienne', *Annales de Géographie*, 84, 1975, pp. 557–88

more than half of all the tractors were in Argentina, Uruguay and southern Brazil. In some regions simple wood ploughs drawn by oxen were in use, but most of Latin America's land was worked only with a hoe, machete, digging stock or foot plough. In 1950 only one-quarter of Brazil's farms had ploughs and human labour was still the major source of power.[20]

There seemed to be straightforward explanations for this low

Table 9.7   Yields of food crops, 1948–50 (100 kilograms per hectare)

	Latin America	Asia	Africa	Europe
Maize	10.7	10.2	7.4	15.5
Wheat	10.6	7.6	6.0	15.8
Rice	16.9	15.8	9.6	42.3
Barley	10.6	10.2	6.7	21.5
Potatoes	52.4	68.9	54.4	163.0

Source: FAO, Production Yearbook 1957, vol. 11, Rome, 1958, pp. 31–2

productivity. On the hacienda only very small proportions of the land were actually used for crops, most being idle, and large amounts were fallow. The landowner could derive from this a more than adequate income, for the labour force was captive, tied to the land by servile tenancies. Labour costs were low, and land untaxed or grossly undertaxed. Many landlords were absentee, and profits either went in conspicuous consumption or were invested elsewhere than in agriculture. On the other hand the bulk of the population was on small farms, frequently fragmented, often on steep slopes, with no access to capital and few resources other than their labour. Although their crop yields were higher than those on the larger farms, there was little prospect of increasing them. For all this, although in 1950 the yield of food crops in Latin America was well below those in Europe, they were very similar to those in Asia and significantly above those of Africa (table 9.7).

THE ADOPTION OF NEW CROP VARIETIES

Although the expansion of area in crops has been the major cause of increased food output in Latin America since 1950, yields have increased; one authority claims that the general level of yields rose 40 per cent between 1950 and 1976. This seems a little high, but there has certainly been a significant increase in the use of modern inputs. The breeding and diffusion of high yield varieties is normally associated with Asia, but the earliest advances were made in the Americas. Before the Second World War hybrid maize was bred in the United States. Combined with the liberal use of fertilizer, but not irrigation water, hybrids had largely replaced open-pollinated varieties in the United States by the 1950s, giving good increases in yields. Hybrids adapted to the climatic conditions of parts of Latin America were bred after the Second World War; but their adoption was slow. In Mexico only one-quarter

Table 9.8 Percentage of sown area in modern varieties, 1982–3

	Wheat	Rice	Maize[a]
Asia (Communist)	31	81	71
Asia (non-Communist)	79	45	36
Near East	31	8	46
Sub-Saharan Africa	51	5	51
Latin America	78	33	54

[a] 1983–6.
Source: M. Lipton and R. Longhurst, New Seeds and Poor People, London, 1989, p. 2

of the area in maize was planted with hybrids by 1970. In Mexico, as in most of Latin America, maize is grown mainly in the traditional, peasant sector and here there was resistance to the adoption of new varieties. Many preferred the taste of the traditional strains, whilst hybrid seed had to be bought each year from a merchant; in contrast the farmer using open-pollinated seed could use part of his harvest to sow next year's crop. However, by the early 1980s about half the maize area of Latin America was sown with hybrids (table 9.8) and yields were estimated to have risen by one-third between 1960 and 1980.[21]

Better known are the semi-dwarf wheat varieties developed by the Ford Foundation in Mexico and released to farmers there in 1961. These varieties (table 9.8) need irrigation and fertilizer for optimum yield, and have been very widely adopted in Asia. But wheat is not grown by irrigation in most of Latin America – north-west Mexico being a notable exception – and is not suited to the climate of most of the continent. Much of the area sown with high yielding varieties is found in Argentina, where it neither is irrigated nor receives much fertilizer. Its virtue there is its short growing season, which allows a soybean crop to follow. Although a high proportion of Latin America's wheat is in new high yield varieties (table 9.8), wheat occupies only a small proportion of the area in food crops. In Mexico, however, new varieties did have a spectacular success. Virtually all wheat was semi-dwarf by 1970, and yield quadrupled between 1950 and 1975. Rice is not a traditional crop in Latin America, and two-thirds is not paddy but upland rice. The area under rice has, however, increased considerably in the last forty years, from about 3 million hectares in 1950 to 8 million in 1987. The introduction of the Philippine varieties into Colombia in the mid 1960s was a spectacular success; local modifications of IR-8 have replaced the indigenous varieties, and irrigated rice the once dominant upland varieties. Rice has become the country's leading food crop. Even so traditional upland varieties remain the main source of rice in most of Latin America (table 9.8), particularly in Brazil, the major producer, where it is mainly grown as a mechanized dry crop. Sorghum is not a major crop in Latin America but improved varieties were introduced to Mexico from Texas after 1955 and form the basis of the country's livestock feed.[22]

Although in places the adoption of new varieties has led to prodigious increases in crop yields, the adoption rate has been very variable. Only one-third of the rice, half of the maize, but four-fifths of the wheat is sown with the new varieties, not always with the benefit of irrigation or fertilizer. This substantial innovation, although more important in Latin America than in Africa or the Near East, has bypassed the majority of Latin America's farmers, a theme which will be discussed later. It must also be recalled that although the adoption of new inputs has occurred in some areas, giving increased yields, in others the reduction of fallowing or overcropping has led to a decline in yields.

FERTILIZERS

In 1950 little chemical fertilizer was used in crop production in Latin America; but nor was it much in use in Africa or Asia at that time. The most liberal usage was on the sugar plantations of Cuba and the Peruvian oases; it was rarely used in Argentina, then technically the most advanced country, and in Brazil 75 per cent of all the small quantity of fertilizer was used in the southern states of São Paulo and Rio Grande do Sul. Since 1950 the consumption of fertilizer per hectare has increased at a considerable rate as it has in all the developing regions (table 8.1); Latin American consumption levels are well above Africa, less than in the Near East and Far East and well below Europe, China and North America. However, the progress of fertilizer consumption per hectare may be gauged by the fact that Latin American levels are now comparable with those of North America in 1960. Although Latin America has to import fertilizer, unlike Africa it has a substantial home industry, producing 64 per cent of its nitrogen fertilizer, two-thirds of its phosphates but less than one-fifth of its potash. But the home production is for the most part at high cost, so that fertilizers are much more expensive per unit of output than in the United States or Western Europe. To buy 1 kilogram of nitrogen fertilizer the United States farmer has to sell 2.4 kilograms of corn, the Chilean farmer 5, the Brazilian 8 and the Uruguayan 10.[23]

Table 9.9   Tractors per 1000 hectares of arable land, 1950 and 1988

	1950	1988	Increase fold
North America	18	22	1.22
Oceania	7	8	1.14
Europe	6	13	12.10
USSR	4	11	2.75
Latin America	1	8	8.00
Far East	0.6	4	6.60
Sub-Sahara Africa	0.5	1	2.00
Near East	0.3	12	40.00
Asian centrally planned	0.01	9	900.00

Sources: FAO, *Production Yearbook 1957*, vol. 11, Rome, 1958; *Production Yearbook 1988*, vol. 42, Rome, 1989

MECHANIZATION

There is no doubt that there has been a substantial adoption of new inputs that raise crop yields in Latin America since 1950, in contrast to much of Africa. There has also been a notable adoption of machinery, whose consequences have not always been happy.

In 1950 most of Latin America's arable land was worked with the hoe. Oxen and the plough – in Mexico the horse and the plough – cultivated an unknown but comparatively small area, and tractors were rare, working perhaps 15 per cent of the continent's cultivated area. Most of the tractors in use were to be found in Argentina, Uruguay, southern Brazil and Mexico, but the number of hectares of arable worked with the tractor was low compared with the developed countries (table 9.9), although substantially above Africa or Asia. Since 1950 there has been a remarkable increase in the number of tractors in use in Latin America, from about 128,000 in 1950 to nearly 1.5 million in 1987. In the same period the arable area has doubled, so the number of tractors per hectare of arable has risen considerably, although at no greater rate in Latin America than in the Near East or China. None the less the area worked by tractor increased to approximately one-third of the cultivated area in 1980. This means, of course, that two-thirds was still worked by hand or by oxen. One recent and problematic estimate suggests that approximately one-fifth of the work done on Latin American farms was undertaken by machinery, one-fifth by draught animals, but over one-half by hand; this indicates a greater degree of mechanization than in Asia or Africa (table 8.2). However, this must be kept in perspective. As late as 1980 72 per cent of all farms in Brazil had neither ploughs nor draught animals, and only 7 per cent had tractors. Mechanization like the use of other new inputs has made little progress on most of the farms.[24]

Although the level of mechanization – as measured by the use of tractors – in Latin America is above that in the Far East or Africa, but not the Near East or China, it is still well below that of the developed countries, or indeed Europe or North America in the 1950s (table 9.9). There are several reasons for this. Although a number of Latin American countries produce tractors and other implements, they are for the most part expensive, whether imported or made at home. In about 1970 a tractor cost the Brazilian farmer – in real purchasing power – about ten times what it cost the United States farmer. Nor is the tractor economic or appropriate in much of Latin America; on the small farms of much of the uplands of Central America or the Andes, not only has population pressure led to the acute subdivision of farms, hardly justifying the use of machinery, but many slopes are too steep for the use of machinery without the risk of soil erosion. Indeed it was said that the highly eroded nature of much of Mexico's farmland forty years ago was due to the replacement of the hoe by the ox and plough. Elsewhere it has been argued that there are too many machines in use. In much of Brazil there is still a rural labour surplus; in the 1960s and 1970s subsidized credit for the purchase of agricultural machinery and rapid inflation made it cheap to acquire machinery, and the labour force was much reduced at a considerable social cost.[25]

PEASANT AND CAPITALIST

It has been argued that Latin American agriculture has split into two distinct sectors in the last thirty years. Whereas in 1950 most of the hacienda were backward, many of these have, particularly since the 1960s, been converted into capitalist enterprises which have aimed at maximizing profits. They have adopted new inputs, greatly increased their output, adopted labour saving machinery, ended the servile tenancy and the allocation of land in return for labour, replaced the tenant with wage labour and made every effort to reduce the labour force. Further, these farms have concentrated upon export crops.

In contrast is the somewhat loosely defined peasant sector. According to E. Ortega this sector employs half the rural population of Latin America, over two-thirds in the Andean republics; they farmed 45 million of the 105 million hectares sown to crops in 1979, although they had only one-third of the total agricultural area. Their farms are small,

averaging 11 hectares of which only 3.3 hectares are in crops; the bulk
of these holdings are very small, 39 per cent being less than 2 hectares.
Most of these peasants still try to produce their own food. But they are
not subsistence farmers, for not only do they sell much of their output,
but they provide a substantial part of total agricultural output, and
particularly food crops for the domestic market. They are not, however,
exclusively food producers. Thus it has been estimated that in Central
America the output of the small-scale intensive sector is 80 per cent
food for the home market and 20 per cent crops for export; in contrast
the large-scale producers export 75 per cent of their output. In Ecuador
small farms produce much of the banana output and 60 per cent of
the cocoa, and in Mexico, where the large private holdings have come
to dominate the market, nearly half the cotton output is produced by
peasants.[26]

The transformation of the former hacienda sector – or part of it –
into modern capitalist agriculture is by no means easy to trace, much
of its history being buried in polemics. The sources of the new entrepre-
neurial class are various. In countries where substantial parts of large
haciendas were expropriated, those remaining have been converted into
efficient estates. This class has been particularly important in Mexico.
In some of the Andean republics land reform proposals promised to
preserve haciendas that were efficiently organized, and this encouraged
improvement. Foreign-owned plantations, such as the banana companies of Central America and the sugar plantations of the Peruvian
coast, have been in the vanguard of technical change. In some Latin
American countries rapid growth of incomes in the towns has provided
good markets. In the environs of cities in southern Brazil small farmers
have become small capitalist producers of vegetables, milk and fruit,
and urban capital has flowed into many farms. The rise of food
processing industries, both home and foreign owned, increased the
efficiency of farming in certain commodities – barley for breweries and
tobacco, for example, in southern Brazil.[27]

The attitude of the state to agriculture in Latin America has been
ambivalent. In the post-war period most Latin American governments
believed prosperity would only come through industrialization and
therefore erected tariff walls to protect the infant consumer goods
industries. However, foreign earnings were needed to purchase capital
goods for these industries, and the only source of earnings was agricultural exports. The fall in demand for these in the 1930s and 1940s had
encouraged many economists to believe that they had little future. In
the event they have remained buoyant in the long run, particularly
where some countries – such as Brazil – have been able to switch to
new products, notably soybeans. Home production also presented a

dilemma, for rapid population growth needed rapid food output; if this was not forthcoming, rising food prices would have led to rapid wage inflation and halted industrial expansion.

Government policies have varied considerably. In Mexico before 1942 the state land reform encouraged the growth of the *ejido* and was opposed to large-scale farming. However when after 1942 attention was directed to industrialization, it was thought that the *ejidos* would be unable to provide a sufficient growth of food or exports for industrialization so the government encouraged the growth of private farming, investing heavily in irrigation in the north west, where land was sold to private owners. In most countries provision of credit, new seeds and extension work are all oriented to the large-scale farmer rather than the peasant. In Brazil agriculture was largely neglected until 1964, since when there has been credit for the purchase of machinery and an attempt to stabilize producer prices and encourage export crops.[28]

Whatever the vagaries of government policies, the adoption of new inputs has tended to be concentrated on particular crops, on larger farms and in specific areas, and the peasant areas, it is claimed, have largely been bypassed. Thus in Brazil the first tractors were used by European immigrants in Rio Grando do Sul; they spread slowly into São Paulo and Minas Gerais. The total number of tractors in Brazil has risen phenomenally from less than 2000 in 1920 to 8327 in 1950 and 156,000 in 1970, and doubled between 1970 and 1980; yet 86 per cent of all were to be found in the south. Although Mexico has had chemical plants producing insecticides since the 1940s, their use in Latin America is confined largely to cotton, a major export crop in several countries. In Mexico the use of high yielding seeds has been limited to the larger farms, and in particular the irrigated farms. Although seeds and fertilizers can be bought in small amounts by peasants, irrigation in Mexico is largely found in the large farm areas of the north west. According to Cynthia Hewitt de Alcantara, because the new high yielding varieties only flourish with irrigation, 80 per cent of Mexico's farmers have been bypassed by Mexico's green revolution. Certainly agricultural output in Mexico is remarkably concentrated, particularly in the irrigated areas. In the early 1970s, 7 per cent of Mexico's farmers produced 45 per cent of the value of output on only 2 per cent of the cropland; in the 1950s 80 per cent of all the *increase* in output came in the irrigated north west. In Bolivia, where smallholders are still a dominant part of the rural population, new inputs have been largely confined to sugar-cane and cotton, in Brazil two-thirds of the fertilizers are used on cotton, sugar-cane and soybeans, and in Ecuador the same proportion goes on sugar-cane, coffee and bananas.[29]

The size of the peasant sector is a matter of debate, depending on what size of farm is defined as peasant or small scale. Undoubtedly it constitutes a substantial proportion of the population of the cultivated area, although not of total agricultural area for much of the land of the larger estates are either idle, in fallow or in grass (table 9.10). Thus in Colombia, Argentina and Ecuador the smallholders produced over half the value of agricultural output in the 1960s, although with less than half the agricultural area. This was because small farmers cultivated a high proportion of their land and farmed it more intensively.[30] But the significance of the smallholders is that they produce a high proportion of domestic food consumption. Thus in Brazil family farms – those without hired labour – produce 80 per cent of the staple foods; half the manioc and beans, the diet of the poor, comes from holdings with less than 20 hectares. In Mexico small farms produce two-thirds of the maize and beans. In Peru small farms, with only 15 per cent of the total agricultural area, produce half the cereals and three-quarters of the root crop. This split is most marked in El Salvador, where the larger farms produce coffee and farms of less than 5 hectares grow 60 per cent of the maize and beans.[31]

It has been argued that the peasant sector, unable to purchase the new inputs that have increased yields in the capitalist sector, has stagnated. This may be true, but it would be wrong to suppose that all small-scale farming has been undynamic. In Ecuador the peasant sector increased output more rapidly than the agricultural economy as a whole between 1965 and 1977, and in Bolivia the peasant sector, which would appear to be the most backward on the continent, increased output at 4.4 per cent per annum between 1950 and 1976. Nor have all small farmers been unenterprising; in Brazil 60 per cent of the soybean crop is produced on family farms. It is true, however, that the peasant sector has made little contribution to livestock output, which in Latin America is carried out predominantly on large units. It is also true that labour productivity is far greater on the large farms than the small; in Colombia the value of output per hectare is 10 per cent greater on small than large holdings, but output per caput on the large farms is ten times greater than on small holdings.[32]

AGRIBUSINESS, EXPORTS AND FOOD SUPPLY

The remarkable increase in population since 1950 has made it difficult for Latin American farmers to maintain the per caput output of food crops, particularly as much of the undoubted improvement in farming has been confined to the capitalist sector that has emerged in the last twenty-five years. Many have argued that this problem has been com-

Table 9.10 Size of the small holding sector in the 1960s

	Percentage of agricultural families			Percentage of land used		Percentage of value of agricultural output	
	Estates	Landless	Smallholders	Estates	Smallholders	Estates	Smallholders
Argentina	5.2	36.3	58.5	51.9	48.1	42.4	57.1
Brazil	14.6	61.9	23.5	93.5	6.5	78.7	21.3
Chile	9.5	49.7	40.8	92.6	7.4	80.0	20.0
Colombia	5.0	24.7	70.3	72.8	27.2	47.8	52.2
Ecuador	2.4	34.5	63.1	64.4	35.6	40.7	59.3
Guatemala	1.6	27.0	71.4	22.3	27.7	56.4	43.6

Source: A. Pearse, 'Subsistence farming is far from dead', Ceres, 2, 1969, pp. 38–43

pounded by the fact that many of the larger farms have concentrated upon export crops, and that export crops have increased at the expense of the food crops. This has been attributed by some not simply to the preference for the more profitable nature of export crops, but also to the increasing control by foreign companies over Latin American farming.

The presence of foreign companies in Latin American agriculture is far from new. British investment in the Argentine beef industry and the Guyanese and other Caribbean sugar plantations has been long established. American companies developed the banana plantations of Costa Rica, Honduras and Guatemala and the sugar plantations of the Peruvian coast. Since the nationalization of sugar plantations in Cuba, Guyana and more recently Peru, foreign companies have been loath to own land and have concentrated more on the processing of food crops, bought from local growers, and the trading of these commodities on the international market. In addition American companies have been important in the establishment of input industries in Mexico and Brazil, making fertilizers, machinery and pesticides. American corporations have also leased or bought areas in the Amazon basin in the last ten years.

Whether export crops – controlled by foreign companies or not – have grown more rapidly than food crops over the last thirty years is difficult to substantiate. Not all apparent export crops are exported; thus one-third of Peru's sugar output is consumed at home. None the less, exports account for a substantial part of Latin American agricultural output, in 1980 17 per cent of the total value. In individual countries exports appear to have increased at the expense of food production. In El Salvador, for example, food output per caput has declined since the 1950s, but agricultural output has kept up with population growth. On the good volcanic soils food crops have been displaced by coffee, and in the coastal areas new land has been devoted not to food crops but to cotton. In Central America as a whole there has been increasing polarization between the small-scale sector producing food crops for home consumption and the larger farms concentrating upon export crops; in some countries exports account for over 70 per cent of total agricultural output. In the drier lowlands of Central America, particularly on the Pacific coastlands, there has been development of cattle ranching aimed at exports for the United States, often at the expense of food production. Nor is it only in Central America that export crops have increased at the expense of food crops. This occurred in Brazil in the 1970s, with the great expansion of soybeans – 0.75 million hectares in 1970 but 8.5 million by 1980 and over 11 million in 1990. In Mexico the proportion of the arable area in basic food crops fell from three-quarters in 1940 to only half in 1980, due not only to the encouragement of export crops but also of feed crops to meet the demand for livestock products from an increasing middle class. In the 1970s food imports began to increase and by 1980 food imports provided 12 per cent of Latin America's food supply.[33]

CONCLUSIONS

Latin America differs from Asia in that much of its increase in food output has come from the expansion of the area in crops. Furthermore, unlike most of the rest of the developing world it has a large-scale capitalist agriculture, utilizing a wide range of manufacturing inputs, increasing yields and often shedding labour as a result of the use of machinery. This was possible because of the existing distribution of land. Large farms have not had to be built up by amalgamation; they already existed. Not all hacienda have of course been transformed in this way. Finally, few countries in Latin America, in spite of the growth of food imports, have food supplies below national requirements.

In Latin America the problem of hunger is not that agriculture has failed to produce sufficient food, but that large sections of the rural population have insufficient income to buy enough food or land to produce it, and that the extraordinary growth of population in the cities has also left many without jobs. In the rural areas the redistribution of land may provide some solution to this problem, although it should be recalled that in some areas, notably in the *altiplano* of Peru and Bolivia, redistribution has not solved the problem; there was not enough land to go round.

10

Asia

The area of Asia is smaller than that of Africa and only a quarter larger than Latin America, but its population – 3112 million in 1990 – greatly exceeds that of the two other continents. Indeed China alone (1139 million) exceeds the combined population of Latin America (448 million) and Africa (642 million), as does South Asia, the former British India (1128 million). The land mass and its huge population show a remarkable internal diversity.

Asia, unlike South America and Africa, lies largely north of the equator. The area with high temperatures, rainfall throughout the year and a natural vegetation of rain forest – rapidly being reduced – lies in the south in Indonesia and parts of mainland South-east Asia. The bulk of South, South-east and East Asia has a climate dominated by the monsoon. Indrafts of hot, maritime air bring heavy rainfall to most areas for a few months in the summer. During the winter high pressures establish themselves over the Eurasian land mass and southward moving air is cool and dry, although only the north of China, the Tibetan plateau and some northern upland areas have winters where low temperatures preclude crop growth. The variability of the summer monsoon, both in the amount and the timing of its onset, means that farming in much of the continent depends upon irrigation to a much greater extent than elsewhere. This is true even of South-west Asia, where the monsoon has little influence and rain falls mainly in the winter.

The topography of Asia is dominated by the mountains that run from Turkey through Iran and in northern India become the Himalayas and northwards the Tibetan plateau. Much of the interior of mainland South-east Asia is mountainous, and southern China is an upland region with many steep slopes. The major areas of settlement are, in contrast to Africa or Latin America, the deltas and alluvial plains of rivers, notably the Hwang-Ho in northern China, the Ganges-Brahmaputra in northern India, Pakistan and Bangladesh, and the deltas of the Mekong, the Red

River, the Chao Phraya and the Irrawady in South-east Asia. In all these areas, except north-west India and northern China, rice is the major food crop, partly because in the flooded deltas of Bangladesh, lower Burma and Thailand few other crops can be grown, but also because the flat land provides an ideal site for the crop, which gives comparatively high yields even with traditional methods. Rice provides half the total calorific intake of the Asian population.

Economically and politically Asia shows great contrasts. China, Vietnam, North Korea and Mongolia all have centrally planned socialist economies, and Burma has a socialist regime although its farming has yet to be collectivized. India alone in Asia has established a form of parliamentary government, and has a mixed economy. Elsewhere market economies predominate, although state direction is often strong. Over the last forty years most Asian countries have had marked economic growth. But in only two areas has this growth lifted net national income per caput above the lowest rungs (table 10.1). First, in South-west Asia petroleum exports have given several countries above average national incomes per caput, notably Saudi Arabia. Second, a number of countries have had a marked development of manufacturing industry. However, these countries – Taiwan, North and South Korea, Hong Kong and Singapore – contain only a small fragment of Asia's population. China, India, Bangladesh and Pakistan all remain very poor, and Indonesia owes its recent rise in income per caput to oil exports over the last two decades (table 10.1). Although in South-west Asia and in limited parts of East Asia the agricultural population has fallen to less than half the labour force, and in spite of the rapid urban growth in many countries, Asia remains essentially rural and agricultural. Outside South-west Asia, only Korea, Malaysia and the Philippines have less then 70 per cent of their populations living in rural areas (table 10.1). Whereas Latin America and Africa have suffered severely from the debt crisis, most Asian countries experienced continued economic growth in the 1980s.

POPULATION GROWTH AND POPULATION DENSITY

Asia, like Latin America and Africa, has experienced a remarkable growth of population since 1950, and rates of increase remain high (table 10.1). However, in contrast to these continents Asia in 1950 already had very high population densities, and its rate of population increase since has been somewhat slower. Indeed in some countries there has been a marked decline in the rate of increase, notably in China. Because

Table 10.1 Economic aspects of Asia

	GNP per caput (US dollars), 1988	Percentage of labour in agriculture, 1988	Percentage of population in towns, 1988	Agriculture as a percentage of all exports	Rate of population growth, 1980–9 (per cent per annum)
Saudi Arabia	6200	41	76	1.6	4.2
Iraq	n.d.	22	73	0.3	3.6
Iran	n.d.	29	54	3.6	3.0
Korea, Republic of	3600	27	69	1.7	1.2
Malaysia	1940	34	41	24.4	2.6
Turkey	1280	50	47	26.1	2.3
Jordan	1500	6	67	9.4	3.7
Korea, Democratic Peoples Republic	n.d.	35	n.d.	6.5	n.d.
Syria	1680	25	51	13.3	3.6
Mongolia	n.d.	32	n.d.	19.9	n.d.
Philippines	630	48	41	17.5	2.5
Thailand	1000	66	21	31.6	1.9
Indonesia	440	50	27	17.2	2.1
Pakistan	350	51	31	27.6	3.2
China	330	69	25[a]	8.9	1.3
Sri Lanka	420	52	21	43.0	1.5
India	340	67	27	16.8	2.2
Vietnam	–	62	n.d.	17.9	2.4
Afghanistan	n.d.	56	n.d.	45.3	n.d.
Burma	n.d.	48	24	17.6	2.1
Nepal	180	92	9	36.5	2.6
Bangladesh	170	70	13	12.1	2.8
Laos	180	72	18	3.6	2.6
Kampuchea	n.d.	71	n.d.	83.0	n.d.

n.d., no data.
[a] The World Bank gives 50 per cent for China. Other authorities have 25 per cent which seems more plausible.
Source: World Bank, *World Development Report 1990*, Washington D.C., 1990

the proportion employed in agriculture remains high, the density of the agricultural population to the arable area is much higher in Asia than in Latin America or Africa (see figure 8.2). This does not mean of course that all parts of Asia have high densities. Much of the arid south west is sparsely populated, as is the cold and dry Tibetan plateau and north-west China. The interior of mainland South-east Asia is sparsely occupied, and densitites are low on some of the islands of the South-east Asian archipelago, notably Kalimantan and New Guinea (figure 8.3).

There are also marked regional variations in density between the settled areas of Asia. Most notable is the contrast between the agricultural density of East Asia and the rest of the continent. Agricultural densities are very high in China, Korea and northern Vietnam and were equally high, before the post-war industrialization, in Japan. Densities in India and Pakistan are less than half those in China; most of South-east Asia – Java is an exception – has densities well below East Asia (table 10.2).

SOME ASPECTS OF ASIAN AGRICULTURE: CROPS AND LIVESTOCK

Asian agriculture has more in common with the traditional agriculture of Europe than that of modern Africa or Latin America. The typical farm is small, operated by the farmer and his family. Much of the produce is consumed on the farms; in the 1970s 80 per cent of Chinese food production did not leave the farms, in the 1980s the value was 50 per cent. In the 1980s only one-third of Indian grain was sold off the farm, and even in a traditional exporter such as Thailand, two-thirds of all rice output is for consumption on the farms.[1] The land is cultivated with a plough drawn by oxen or water buffalo, although in very densely populated areas only the hoe is used. Most of the seed is broadcast, although in some rice areas, particularly in East Asia, rice is transplanted from nursery fields to the paddies (see table 10.8 later). The sickle is the usual implement for harvesting, although in Java and Thailand a hand knife was used until recently. A great variety of food crops is grown, but rice is dominant, taking second place to wheat only in northern South Asia and northern China. Maize is important in China and parts of South-east Asia. Millets and sorghum are grown in the drier areas which lack irrigation, in northern China and the interior of the Indian subcontinent, but have declined in importance in the last twenty years. Minor food crops such as cassava, peanuts and sweet potatoes are locally important (table 10.3).

Table 10.2 Some characteristics of modern Asian agriculture, c.1986

	Arable as a percentage of total area	Irrigated as a percentage of arable	Agricultural population per hectare of arable	Tractors per thousand hectares of arable	Fertilizer per hectare of arable (kg)	Rice yield (tonnes per hectare)	Wheat yield (tonnes per hectare)	HYVs as a percentage of rice	HYVs as a percentage of wheat	Multiple cropping index
Korea, Democratic Peoples Republic	20	48	3.2	31	331	7.2	4.0	–	–	–
Japan	13	62	2.0	338	427	5.8	3.6	–	–	–
Korea, Republic of	22	58	5.2	8	385	6.7	3.0	47	–	–
China	10	46	7.6	9	174	5.3	3.0	80	25	150
Indonesia	12	34	3.8	1	98	4.1	–	41	–	–
Malaysia	13	8	1.2	3	157	2.6	–	37	–	–
Burma	15	11	1.9	1	21	2.9	1.8	7	–	111
Sri Lanka	29	32	4.6	15	102	3.0	–	63	–	121
Pakistan	27	77	2.7	8	94	2.3	1.7	40	76	136
Philippines	27	18	3.4	2	43	2.7	–	68	–	–
Vietnam	21	26	5.7	6	62	2.7	–	–	–	–
Afghanistan	12	33	1.2	–	11	2.2	1.2	–	–	–
Bangladesh	68	23	8.1	1	67	2.2	1.8	14	73	148
India	57	26	2.9	4	57	2.5	2.0	36	72	120
Nepal	17	28	6.7	1	20	2.1	1.2	18	73	117
Thailand	39	20	1.7	7	24	2.0	–	11	–	1
Laos	4	13	3.4	1	–	1.6	–	–	–	0
Kampuchea	17	3	1.8	–	–	1.2	–	–	–	–

HYV, high yielding variety.

Sources: FAO, *The State of Food and Agriculture 1989*, Rome, 1989; *Production Yearbook 1988*, vol. 42, Rome, 1989; D. Dalrymple, *Development and Spread of High-Yielding Varieties of Wheat and Rice in the Less Developed Nations*, US Department of Agriculture, Washington, D.C., 1978; D. Dalrymple, *Survey of Multiple Cropping in Less Developed Nations*, US Department of Agriculture, Washington, D.C., 1971.

Table 10.3   Land use in Asia, 1961–5 and 1990 (million hectares)

	1961–5	1990
Cereals	294.0	310.3
Rice	114.1	131.4
Wheat	61.9	84.4
Maize	22.4	39.9
Millet and sorghum	68.5	39.3
Roots and tubers	18.4	19.3
Pulses	37.8	35.9
Oilseeds	ˎ 31.1	48.8
Sugar-cane and sugar-beet	4.5	8.4
Coffee	0.48	1.3
Tea	1.1	2.3
Jute	2.5	2.3
Rubber	4.25	–
Cotton	11.7	–
Tobacco	1.9	3.2

Sources: FAO, *Production Yearbook 1989*, vol. 43, Rome, 1990; *Production Yearbook 1976*, vol. 30, Rome, 1977; *Production Yearbook 1965*, vol. 19, Rome, 1966

Oilseeds and pulses make up the bulk of the remaining cropland. Non-food crops constitute a small proportion of the total arable, which is surprising considering the importance of plantation crops in the economies of some countries. Agricultural products account for one-quarter or more of the exports of Malaysia, Pakistan, Kampuchea, Thailand, Sri Lanka and Nepal (table 10.1). In the nineteenth century the development of jute, coffee, tea and rubber in Bangladesh, India, Ceylon and Malaya was almost entirely for export to Europe; in Indonesia sugar-cane, rubber and coffee were grown for the same market, and from the deltas of Burma, Thailand and southern Vietnam rice was exported to other Asian countries as well as to Europe. Agricultural exports remain an important part of the exports of many Asian countries (table 10.1), but a dual system of a peasant sector and a commercial exporting sector occurs only in Sri Lanka and Malaya, perhaps in Sumatra. Elsewhere food crops dominate. Thus even in Bangladesh, where jute provided nine-tenths of all agricultural exports by value, the crop occupies only 7 per cent of the arable area, and in Java the estates occupy much the same proportion.

Livestock are found throughout Asia and their densities are often surprisingly high, considering the competition for land between man and animal. Two livestock enterprises are unimportant in Asia. The extensive rearing of cattle and sheep, which is widespread in Latin America and Africa, is confined in Asia to northern China and Mongolia. Nor is dairying, a major part of the livestock output of developed countries, of much significance. Dairy products are not consumed by the Chinese and certain other peoples, such as the Thai, possibly because of their difficulties in absorbing lactose in the intestine. In India and the Indian influenced areas of South East Asia milking is practised, but the poor quality of both the cows and the fodder, together with the low incomes of the population, make it a minor part of livestock production, although the Indian government has, since the 1960s, encouraged the improvement of breeds and methods. India has 15 per cent of the world's cattle, yet it produces only 0.5 per cent of the annual beef output. Little land is available for feeding livestock in India, which live largely from the straw of food grains and waste land. Cattle thus use most of their feed to fuel their exertions pulling ploughs and wagons, and put on weight very slowly. The Hindu reluctance to slaughter cattle inhibits the development of a livestock industry, and cattle are valued for their draught power and as a source of manure both for the fields and for fuel; this reluctance to slaughter livestock is also found among the Buddhist populations of South-east Asia. The poverty of tropical grasses also hampers development. In much of East Asia and Java the very high density of population excludes the keeping of cattle except – along with water buffaloes – as draught animals, and in China poultry and particularly pigs provide nearly all of the little meat that is eaten; both poultry and pigs are non-ruminants and compete with man for grain and roots, but can utilize food wastes. Over much of Asia – the South-west, Pakistan, Bangladesh, Malaya and Indonesia – few pigs are kept because the populations are predominantly Muslim.[2]

FARM SIZE AND LAND OWNERSHIP

In Africa the dominant production unit is small, and large farms occupy little of the farmland; in Latin America, although the occupiers of very small holdings dominate farm structure, large and very large farms make up most of the farmland. But in Asia farms are very small, *and* they occupy most of the cropland. Nor has there been any tradition of large farms. There has been no significant European settlement and little land appropriation, even though much of the continent was part of European empires until 1948. Plantations were established by the British, Dutch and French, but most of these have been nationalized since independence, and in some cases subdivided; however, even when not subdivided they occupied – and occupy – a small fraction of the agricultural land (table 10.4).

Most farms in Asia are very small. Between one-half and nine-tenths are less than 2 hectares. In Indonesia 70 per cent are less than 1 hectare (table 10.5). Indeed it is debatable where the line is to be drawn between the rural household which is landless and that which has some land and is recorded as a farm; certainly in Java and many other parts of Asia farm families gain a significant part of their income from activities off the farm. The large farm in Asia is normally defined as over 10 hectares; they are a small proportion of all farms, but in some countries, notably in India, Pakistan and the Phillippines, they occupy one-third or more of the farmland (table 10.4). Elsewhere – particularly in Korea, Taiwan, Thailand, Bangladesh and Indonesia – the small farms occupy most of the farmland. In contrast to Latin America, where the most rapid agricultural growth has occurred on large farms, some of the most rapid growth rates in Asia have occurred in countries – Taiwan, Malaya and Korea – where the average size of the farm is below 2 hectares and there are few large farms.

Until 1948 China was a country of small farms, many of them rented to tenants. Between 1948 and 1952 the land of landlords and some of the richer peasants was seized and distributed to the landless and those with smaller holdings. China thus became a country of small owner-occupied farms, similar to much of the rest of Asia, although with less tenancy (table 10.6). However, in the period 1952–8 the small family holdings of the Chinese peasant were grouped into successively larger units; first were the mutual aid teams, then the agricultural production co-operatives, the advanced agricultural co-operatives and in 1958 the communes. Private ownership of land was abolished in 1956–7. The level of decision making and accounting varied; at times decisions were made at commune level, which may have consisted of 5000 households, at other times decisions were made at village level. After 1977 and the introduction of the production responsibility system, the accounting unit has shifted back towards the village and household level; between 1979 and 1983 households became the effective farm unit in China again, and in 1985 the communes were abolished, leaving 180 million farms averaging 0.5 hectares in size.[3]

In the socialist countries of Asia land is not owned by individuals; even in China, where most features of a socialist agriculture have been dismantled, the State still owns the land. Elsewhere the bulk of the agricultural land is farmed by owner occupiers, but tenancy does persist and is regarded as a problem. Tenant farmers rarely have the security that they possess in most European countries, and may be evicted at short notice. However, landlords in most parts of Asia are not like those

Table 10.4 Farm size in Asian countries, 1970s

	Number of farms (%)				Area of farmland (%)			
	0–2 hectares	2–5 hectares	5–10 hectares	10 hectares and over	0–2 hectares	2–5 hectares	5–10 hectares	10 hectares and over
Bangladesh	88	10	–	2	58	32	–	10
India	70	18	8	4	21	26	23	30
South Korea	94	–	–	–	80	–	–	–
Malaya	72	27	–	–	48	47	–	–
Pakistan	50	27	15	8	9	23	26	43
Philippines	41	40	13	6	11	31	24	33
Thailand	50	30	16	4	20	32	32	16
Nepal	87	10	2	1	46	31	12	11

Sources: Asian Development Bank, *Rural Asia: Challenge and Opportunity*, New York, 1977, p. 98; *Economic Bulletin for Asia and the Pacific*, **30** (1), 1979, pp. 24–53

Table 10.5 Farm size in selected Asian countries (number of farms in each size group, as a percentage of all farms)

	0–0.5 hectares	0.5–1.0 hectares	1–5 hectares	5–10 hectares	10 hectares and over	Average size (hectares)
Sri Lanka	35.3	30.0	30.2	3.2	1.2	1.38
India	19.4	–	43.5	18.9	18.2	2.3
Philippines	4.1	7.4	69.5	13.4	5.6	2.58
Thailand	10.3	8.3	56.9	19.2	5.4	2.95
South Korea	35.0	32.0	32.0	1.0	0.0	0.91
Indonesia	45.6	24.7	27.5	1.6	0.6	0.99

Source: R. Montgomery and Toto Sugito, 'Changes in the structure of farms and farming in Indonesia between censuses, 1963–1973: the issues of inequality and near landlessness', *Journal of South East Asian Studies,* **11**, 1980, pp. 348–65

Table 10.6 Farm size in China after the land reforms, 1948–52

Class	Percentage of families	Percentage of total area	Average area owned (hectares)
Poor peasant	57.1	46.8	0.8
Middle peasant	35.8	44.8	1.3
Rich peasant	3.6	6.4	1.75
Landlord	2.6	2.1	0.8

Source: Azizur Rahman Khan, 'The distribution of income in rural China', in International Labour Office, *Poverty and Landlessness in Rural Asia*, Geneva, 1977

Table 10.7 Tenants and owner occupiers in selected Asian countries (percentage of holdings)

	Fully owned by operator	Partly owned by operator	Tenant (cash and share crop)
Sri Lanka (1947)	67.0	–	33.0
India (1960)	60.2	22.9	16.9
Philippines (1960)	44.7	14.4	39.9
Thailand (1963)	81.9	14.0	4.1
South Korea (1969)	73.5	19.6	7.0
Indonesia (1973)	74.8	22.0	3.2
Pakistan (1960)	41.0	17.0	42.0

Sources: R. Montgomery and Toto Sugito, 'Changes in the structure of farms and farming in Indonesia between censuses, 1963–1973: the issues of inequality and near landlessness', *Journal of South East Asian Studies*, 11, 1980, pp. 348–65; Azizur Rahman Khan, 'Poverty and inequality in rural Bangladesh', in International Labour Office, *Poverty and landlessness in Rural Asia*, Geneva, 1977, pp. 137–60

of Latin America, owning vast estates; more commonly landlords have comparatively small amounts of land divided among tenants – often share croppers – in one or a few adjacent villages. Information on the extent of tenancy is often out of date and difficult to interpret. For those countries for which data exist, however, owner occupiers are a majority everywhere except in the Philippines and Pakistan (table 10.7). About 43 per cent of the farmland in Pakistan is farmed by tenants, 20 per cent in India and 23 per cent in Bangladesh.[4]

Land reform has been urged in Asia on grounds of both equity and efficiency, but outside East Asia much has been proposed but little done. In Taiwan and South Korea reforms first controlled rents, then transferred ownership to the occupier and placed ceilings upon the amount that could be owned or farmed. There is thus a remarkable equality in farm size and little tenancy.[5] The transfer of land in China to the poorer peasants involved 46 per cent of the arable, and was thus the most comprehensive reform in Asia, although all peasants subsequently lost ownership in the collectivization of 1956–8. Several states have attempted to redistribute land by putting a ceiling upon the amount an individual could own, and some land has been expropriated and redistributed. The amounts, however, have generally been small, for the laws on redistribution have proved all too easy to evade.[6] However, in some countries land redistribution, even if fully accomplished, could do little to solve the problems of landlessness and near landlessness. In Java in 1960, in an attempt to redistribute land declared surplus, 350,000 hectares were put above the ceiling of permissible ownership. However, there were at the time 3 million families seeking land.[7]

Although it has been claimed that in many parts of Asia tenants are being evicted and larger holdings are formed under the spur of commercialization and mechanization, reliable figures on farm size are few and far between, and suggest that it is the increase in the number of small farms that is most noticeable. This is not surprising, given the formidable increase in the farm populations over the last forty years, and the custom of dividing land among all the sons – and often daughters as well – at inheritance.

ASIAN AGRICULTURE IN 1950

Long before the post-war decline in mortality and the subsequent rapid growth of population, much of Asia had very high population densities; there was a notable difference in densities between East Asia and the rest of Asia. Agricultural technologies were largely traditional and farm work depended almost exclusively upon human and animal labour; few if any inputs were bought. East Asia and the Indian subcontinent had possessed high agricultural densities for centuries, and had adapted to the slow growth of population by increasing labour inputs. This intensity was greatest in south China, Japan, Korea and north Vietnam, and was markedly lower in South-east Asia, where the rice exports of the deltas of Burma, Thailand and south Vietnam were a comparatively recent development. Thus over long periods, and because of the lack of easily cultivable land, the populations of East Asia had increased rice yields by using more and more intensive practices. Wet rice needs very large inputs of labour. The crop is grown with the stalk partially submerged for much of the growing season, and thus has to be grown in small, flat fields surrounded by bunds, together with some means of withdrawing water before the crop ripens and is harvested. Most rice is *not* irrigated; two-thirds of the rice grown in South and South-east Asia relies upon the monsoon rainfall, either directly or by flooding from rivers.[8] If the land is irrigated, further water control measures are needed, requiring a considerable expenditure of labour and capital. Rice can be broadcast in the paddy fields, or the seed can be sown in nursery plots and transplanted into the main fields after several weeks growth. As in the provision of irrigation, this increases yields. The frequency of weeding can also be increased to raise yields. Rice benefits from the blue-green algae that live in the flooded paddy fields and add nitrogen to the soil. Livestock manure is scarce in much of Asia because of the shortage of fodder, but in China and certain other East Asian countries night soil was applied to the rice crop. If the land was irrigated – or if the monsoon was particularly heavy – the rice harvest could be followed by another cereal crop, exceptionally by a second crop of rice (table 10.8). Thus by 1950 – indeed by a much earlier date – there was a marked difference in rice yields between East Asia and the rest of the continent, reflecting greater labour intensity which in turn roughly reflected differences in agricultural population density.

In 1950 Japan had the highest rice yields in Asia, a result of half a century of technical change that foreshadowed the Green Revolution that began elsewhere in Asia in the 1960s. In the 1890s most of Japan's rice was already irrigated. Careful selection of indigenous varieties of rice produced one – the *ponlai* – that responded well to chemical fertilizers, and the combination of an improved variety, irrigation and increased fertilizer application increased yields between 1890 and 1910. Japan acquired Korea and Taiwan as colonies in 1900, and used these countries as a source of rice. In both countries the Japanese increased the area under irrigation, began extension schemes, aranged credit schemes for farmers and provided chemical fertilizers. Rice output increased markedly between 1920 and 1940, partly due to higher yields but also due to an increase in the area double cropped, which the extension of irrigation made possible. Thus by 1950 there was a pronounced difference in rice yields between Japan, Taiwan and Korea, and the rest of Asia; in south China comparable yields were due to the great amounts of pig, draught animal and human excreta used as organic manure, a practice uncommon elsewhere in Asia.[9]

THE GROWTH OF FOOD OUTPUT, 1950–80

Little is known of the rate at which agricultural production increased in Asia before 1950. It has been suggested, however, that in China food

Table 10.8 Rice production in Asia, c.1950

	Yield (tonnes per hectare)	Percentage of rice area irrigated	Percentage of rice transplanted	Percentage of land double cropped	Percentage of land double cropped with rice	Chemical fertilizer (kilograms per hectare of arable)
Japan	5.1	96	95	35	0.3	146
South China	4.5	62	–	66	13–15	4[a]
Taiwan	3.2	79	–	93	42	88
South Korea	3.3	58	100	28	–	53
Tongking Delta	2.5	46	–	–	50	–
Mekong Delta	2.1	20	80	low	10	–
Thailand	1.5	24	80	low	–	0.3
Cambodia	1.2	–	–	–	–	–
Lower Burma	1.6	11	90	low	–	0.02
Malaya	2.4	11	94	low	6	0.4
Java	1.7	49	79	25	15–20	1.0[b]
Philippines	1.2	30	80	–	16	0.5
Bangladesh	1.7	12	74	24	–	0.2[c]
Sri Lanka	1.7	60	6	–	32	–
India	1.5	20	–	13	–	0.5

[a] All China.
[b] Indonesia.
[c] All Pakistan.
Source: D. B. Grigg, The Agricultural Systems of the World: An Evolutionary Approach, Cambridge, 1974, p. 78

Table 10.9 Food grain output, India and China (million tonnes)

	India	China
c.1950	60	163
1964–9	90	194
1969–71	118	243
1981–2	134	318
1983–4	152	407

Sources: B. H. Farmer, *An Introduction to South Asia*, London, 1984, pp. 174–5; A. Doak Barnett, *China and the World Food System*, London, 1979, p. 37; S. Ishikawa, 'China's economic growth since 1949: an assessment', *China Quarterly*, **94**, 1983, pp. 217–18; G. Etienne, *Food and Poverty: India's Half Won Battle*, London, 1988, p. 195; Yeh Chi, 'Mainland China's grain crisis', *Issues and Studies*, **25**, 1989, pp. 68–82

output just about kept pace with population growth from the 1880s to the 1930s; in India it did not, and food production per caput fell from 1900 to 1947. In contrast many countries in South-east Asia had rapid rates of increase in agricultural output from the late nineteenth century until the 1930s, but much of this was not of food for home consumption but crops grown for export – rubber, rice and sugar-cane.[10]

Since 1950 the rate of increase in food production has risen well above the level achieved before the Second World War, and in most countries has been sustained over four decades. Some rates of increase, even before the introduction of new high yield varieties, have been very impressive. In the 1950s Punjab output was increasing at 5.5 per cent per annum, whilst in Taiwan and the Philippines agricultural output rose at 4 per cent per annum. The two most populous countries, India and China, could not match these figures, but have substantially increased their grain output since 1950[11] (table 10.9).

There have obviously been variations between countries and between decades, but with the exception of the Near East the rate of increase has not slackened, and in contrast to Latin America and Africa food output was still growing rapidly in the 1980s (table 10.10). Indonesia and China exhibit among the most spectacular recent increases in output. The Indonesian government utilized part of its oil revenues to introduce the new high yielding varieties (HYVs) in the 1970s and 1980s. In China dramatic increases in food output occurred after 1978. Until that time Chinese agriculture had a socialist or command agricultural economy. Large units of production, the communes, were required to deliver quotas of grain and other products to the state, and commune members had little incentive to work hard or take risks.

Table 10.10   FAO indices of food production and food production per caput in the 1960s, 1970s and 1980s

	1961–5 to 1969–70[a]	1969–71 to 1979–80[b]	1979–81 to 1988–9[c]
The growth of food output			
Near East	121	136	122
Far East	121	131	134
India	118	124	141
Pakistan	128	136	142
Indonesia	118	139	145
Asian CPEs	n.d.	136	142
The growth of food output per capita			
Near East	102	105	94
Far East	103	107	111
India	100	100	117
Pakistan	106	104	104
Indonesia	101	115	124
Asian CPEs	n.d.	109	126

n.d., no data; CPEs, centrally planned economies.
[a] 1961–5 = 100.
[b] 1969–71 = 100.
[c] 1979–81 = 100.
Source: FAO, *Production Yearbooks*, Rome, various dates

However, between 1979 and 1985 the system of farm organization was radically changed. The household had replaced the commune as the production unit, the policy of regional self-sufficiency was ended, allowing specialization according to comparative advantage, prices were increased and local markets encouraged. Once Chinese peasants could see some reward for individual effort, output greatly increased. Between 1978 and 1984 grain output increased by one-third to a record harvest of 407 million tonnes, cotton output tripled and meat production doubled.[12]

However, food output has to be related to the increase in population and then the performance seems less impressive (table 10.10). In the 1960s food output just kept pace with population growth, and marked increases in output per capita in the 1970s were confined to a few countries, but by the 1980s substantial increases were widespread except in the Near East where output per capita declined. In India food output just kept ahead of population growth from the 1950s to the 1970s, but in the 1980s has made significant advances. Much the same

Table 10.11 Percentage of cereal output increase due to higher yields

	1961–5 to 1970	1970–9
South and South-east Asia	66	72
East Asia	73	42[a]
South-west Asia	37	65
Africa	31	18
Latin America	37	54

[a] The low figure is probably due to the increase in double cropping.
Source: FAO, *The State of Food and Agriculture, 1979*, Rome, 1980, pp. 6–16

is true of China, where there was little change in output per capita between the 1950s and the 1970s, but between 1978 and 1985 there was a substantial increase in output per capita and food availability per capita.[13] By the mid 1980s, most countries in Asia had a supply of food above national requirements (figure 2.4).

YIELD AND AREA

The long history of settlement in much of Asia meant that by 1950 there was little cultivable land left unoccupied, and most of the arable land supported dense agricultural populations. Some land colonization has taken place in Asia since the war (see pp. 103–5), and in some countries extra food output has come primarily from area expansion rather than yield increases, notably in Thailand. But, as in Latin America, the proportion of increased output attributable to extra area has declined since 1950. In India approximately half the increase in grain output in 1950–60 was due to area expansion, but between 1960 and 1980 80 per cent of the increase was due to higher yields. In China the arable area is less now than in 1950, although the sown area has increased substantially because of the spread of double cropping; none the less it has been estimated that 90 per cent of the increased food output since 1950 has been due to higher yields.[14]

In the 1960s, 1970s and 1980s yield increases were the most important source of increased output in all parts of Asia (table 10.11) and were far more important than in Latin America or Africa. This is not surprising. Most estimates of the potential arable area in Asia suggest that the land remaining uncultivated which could be used for crops is no more than one-fifth of the present arable area, whereas in Latin America and Africa the potential arable land greatly exceeds that in use (see table 6.4).[15] Area expansion can be achieved by colonizing new land, by reducing the fallow or by multiple cropping. The last will be dealt with later; land colonization has already been touched upon (see pp. 103–5). Something must be said of fallow reduction.

In Africa and Latin America large areas of arable land were in fallow in the 1960s, and in Africa the reduction of this fallow has been an important source of increased food output. In Asia virtually all the arable is sown each year. However, this figure is a composite; in most Asian countries some of the arable is sown more than once each year, and a much smaller amount is farmed by shifting cultivation and is sown only once in several years. Most of the shifting cultivation remaining in Asia is to be found in the interior of mainland South-east Asia, or in Sumatra, Kalimantan or Papua New Guinea. In Indonesia both food and cash crops are grown in *ladang*; rubber is particularly easy to integrate into *ladang*. But shifting cultivation supports a comparatively small proportion of Asia's population, and what reduction of fallow has taken place has added little to food output in Asia in the last forty years.[16]

CROP YIELDS IN ASIA SINCE 1950: THE GREEN REVOLUTION

In the mid 1960s HYVs of wheat were introduced into Asia from Mexico, and a high yielding rice, 1R-8, was developed at the International Rice Research Institute (IRRI) in the Philippines and made available in the rest of Asia. Together with the use of chemical fertilizers, pesticides and irrigation, these crops gave yields well above those of traditional varieties. The adoption of these new varieties was rapid; the consequences of these changes (unfortunately dubbed the Green Revolution) have given rise to a large and often polemical literature. The yields obtained by farmers have been lower than those obtained on experimental farms, a fact that would not surprise most agronomists. The use of the 'miracle' seeds has not solved the problem of hunger in Asia, if for no other reason than that the population of Asia increased by 1200 million between 1965 and 1989; yet the great increases in output which have resulted undoubtedly prevented many deaths that would have occurred without this increase. The new seeds have been widely adopted by those growing wheat and rice (see table 10.16 later); but wheat and rice, although the major food grains, are not all the food supplies of Asia. Thus in 1980 only about one-quarter of the seed used for all food grains in South and South-east Asia consisted of new HYVs.[17] As a revolution the changes are clearly less than a total transformation of rural Asia. But the rapid growth of output produced on those farms where the new technology has been adopted has helped to maintain the rate of increase of food output and has prevented price rises that would have occurred in the absence of increased output. Some critics, who have reluctantly agreed that there have been improvements in food output and crop yields, have argued that the adoption of the new technologies has benefited large farmers and not small, increased the number of landless and lowered their incomes, caused the eviction of tenants and increased the gap in incomes between the landless and the small farmers on the one hand and the large farmers on the other. The literature on these subjects is confusing and controversial, and varies from the broad generalization applied to the whole of Asia, with little reference to any substantial facts, to the careful study of two or three villages over two or three years. The problems are compounded by the difficulty of obtaining reliable information on incomes, farm size and landlessness; further, even where it can be shown that farms are getting larger, landlessness is increasing or agricultural wages are falling, these can often be plausibly explained by causes other than the spread of a new technology. The growth of population, for example, can surely be a possible cause of both increasing numbers without land and falling agricultural wages. These matters have been much discussed.[18] The purpose of the following sections is only to assess the impact of the new farming methods upon food output.

There have been great variations in the increase in crop yields in Asia. Thus in Pakistan the adoption of semi-dwarf wheats led to a doubling of yields between 1965 and 1976. In the Indian Punjab the new varieties led to increases in rice yields of 174 per cent between 1965 and 1981, whilst in Sri Lanka rice yields rose 50 per cent in the 1970s. On the other hand rice yields have increased very slowly in Burma and Thailand over the last thirty years.[19] Furthermore, so much attention has been paid to the period after 1965 when the first new varieties were introduced that it has been forgotten that yield increases were being obtained before 1965 without the use of the high yielding varieties. Indeed, in Taiwan, Korean and Japan rapid increases in yield were obtained before 1950 by using more fertilizer and improved varieties, by increasing the irrigated area and by more careful cultivation.

A comparison of the average yields of the major food grains in Asia for 1950–9 and 1980–9 allows a more realistic appreciation of crop yield increases (table 10.12). The greatest increases came in wheat, which is the second ranking grain in Asia, and the least in millets. Data for China, which probably exaggerates yields and their increase, suggest a similar pattern in that country (table 10.13).[20] Rice yields increased less than either wheat or maize.

Table 10.12 Average crop yields in Asia, excluding China and Japan (tonnes per hectare)

	1950-9	1960-9	1970-9	1980-9	Percentage increase
Wheat	0.83	0.94	1.28	1.6	93
Rice	1.39	1.66	1.99	2.3	65
Maize	0.92	1.12	1.28	1.6	74
Millets and sorghum	0.44	0.48	0.57	0.7	59

Source: FAO, *The State of Food and Agriculture 1980*, Rome, 1981, p. 17; *Production Yearbook 1989*, vol. 43, Rome, 1990

Table 10.13 Crop yields in China, 1952–89 (tonnes per hectare)

	1952-7	1988-9	Percentage change
Rice	2.5	5.4	116
Wheat	0.83	3.0	217
Maize	1.34	3.8	183
Millet	1.1	1.9	73
Gaoliang	1.2	–	–

Source: K. R. Walker, 'China's grain production 1975–80 and 1952–7: some basic statistics', *China Quarterly*, 86, 1981, p. 223; FAO, *Production Yearbook 1989*, vol. 43, Rome, 1990

NEW CROP VARIETIES

In India some 40,000 varieties of rice are grown;[21] other food grains also have numerous varieties. These have been selected by farmers over very long periods to adapt to local microenvironments. Thus in the flooded deltas of the Irrawady and Chao Phraya floating rice varieties are used, long-stalked rice that can survive the rapid rise in the water level during the monsoon floods. In the more northerly locations, such as Japan and Korea, there is a premium on rapid maturity; the indigenous varieties of Japan also have short stalks that prevent lodging. Varieties also have to be adapted to changes in day length, which varies with latitude. Local varieties often have a built-in immunity to local disease. There are also varieties suited to rain-fed conditions and others to the more regular water supply of irrigated areas. Most traditional varieties are adapted to give reasonable yields under poor or uncertain ecological conditions rather than to give high yields under optimum conditions. Up to the post-war period improved varieties were selected either by farmers from observation of their better yielding crops or, more exceptionally, by agricultural scientists from observation of many varieties grown for comparison on experimental farms; this was the case with *ponlai* rice in Japan, later taken to Taiwan and Korea. They gave yields 20–30 per cent above the indigenous varieties in the 1930s.[22]

This type of selection proceeded in many parts of Asia, especially in India, after 1950. However, the most notable advances came in the 1960s in China and the Philippines. Advances in plant breeding had allowed the selection of specific qualities in seed. In the early 1960s a rice variety IR-8 was bred at the IRRI in the Philippines which had a number of characteristics. First, it was not sensitive to changes in day length and could be grown at any latitude. Second, it had a short stalk and did not lodge with a large grain. Third, it was more responsive to chemical fertilizer than any traditional variety then grown. This – and later rice varieties developed from IR-8 – gave much higher yields than any traditional varieties. However, they needed a controlled water supply, which meant irrigation; a far greater dosage of fertilizer than was being used anywhere in Asia except Taiwan, Japan and South Korea (table 10.17); and, because they were thought to lack the immunity to disease that the indigenous varieties possessed, the use of pesticides. In 1965 Mexican semi-dwarf varieties of wheat were introduced into India and Pakistan; they also gave much higher yields than traditional varieties if provided with an adequate moisture supply, chemical fertilizers and pest control. In the 1960s the Chinese produced, independently of the research in the Philippines, a dwarf indica rice that required chemical fertilizers and a controlled water supply; it gave yield increases comparable with IR-8 but matured more rapidly. Mexican wheats were imported into China in 1969 and cross bred with local varieties.[23] Since the introduction of the first IRRI varieties in 1965–6, the original varieties have been improved and, most important, adapted by cross breeding to local environments. Indeed most improved varieties – the term that has recently replaced HYVs – are now bred in the countries where they are used. Not only have further increases in yield between obtained, but immunity to diesease has been developed. Modern varieties are no more susceptible to disease than traditional varieties. However comparatively little progress has been made in breeding varieties suitable for semi-arid or flooded conditions.[24]

The new varieties were rapidly adopted by farmers. By 1976–7 nearly half the wheat and rice in Asia was sown with HYVs or the improved Chinese varieties (table 10.14). But there was a great difference between China and the rest of Asia. Approximately 80 per cent of the rice in China was sown with improved varieties, but in the rest of Asia

Table 10.14 High yielding varieties and improved Chinese varieties: area sown to rice and wheat, 1976–7

	Rice		Wheat	
	Area in HYVs (million hectares)	HYVs as percentage of all rice	Area in HYVs (million hectares)	HYVs as percentage of all wheat
China	28.96	80	7.03	25
Rest of Asia	24.2	30.4	19.67	72.4
All Asia	53.16	45.7	26.70	48.4

HYV, high yielding variety.
Sources: D. G. Dalrymple, *Development and Spread of High Yielding Varieties of Wheat and Rice in the Less Developed Nations*, Washington, D.C., 1978; R. C. Hsu, *Food for One Billion: China's Agriculture since 1949*, Boulder, Colo., 1982, p. 65; R. Barker, D. G. Sisler and B. Rose, 'Prospects for growth in grain production', in R. Barker and R. Sinha (eds), *The Chinese Agricultural Economy*, London, 1982, pp. 163–81

only 30 per cent was sown with IRRI rices. Conversely, whereas only 25 per cent of the wheat was sown with new varieties in China, some 72 per cent of the wheat area in the rest of Asia was sown with Mexican wheats. This is explained by state policies and the existence of irrigation facilities. Prior to the introduction of the IRRI rice varieties in India the government was aiming to concentrate improvements in the areas with good irrigation by providing improved – but not high yielding – varieties, fertilizers and credit. Not surprisingly, when the new wheat and rice became available attention was first turned to the irrigated areas, for both the wheat and rice HYVs give their best yields under irrigated conditions. But only 36 per cent of the wet rice in Asia (excluding China) is irrigated. Elsewhere the crop relies upon the monsoon rains, which can be very variable in both quantity and timing. Furthermore the short stalked IRRI varieties cannot be grown in rice areas that are flooded during the monsoon.[25] Thus they could not be easily grown in the delta of the Ganges-Brahmaputra (except under irrigation in the dry season), the Irrawady or the Chao Phraya. Indeed the wide differences in the adoption rate of HYV rice could be directly related to the proportion of the area of rice that can be irrigated (figure 10.1 and table 10.15). Very little of the rice area of Burma, Thailand or Bangladesh was – or is – sown with the new varieties. Most of the wheat sown in Asia outside China was grown in northern South Asia in relatively dry areas with, however, well-developed irrigation systems. This zone

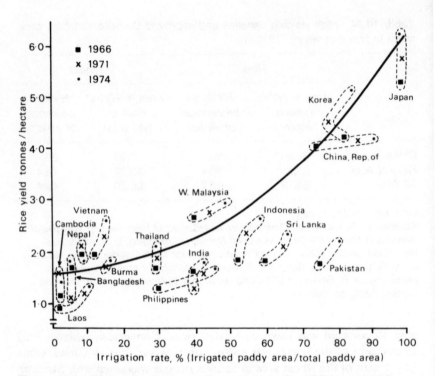

Figure 10.1 Rice yields and irrigation in Asia in the 1960s and 1970s
Source: Asian Development Bank, *Rural Asia, Challenge and Opportunities*, New York, 1978, p. 157

includes the Punjab region in Pakistan and India, Haryana and western Uttar Pradesh, and here there has been a very rapid and successful adoption of HYV wheat.

In China the situation was quite the opposite. In the late 1950s, before the introduction of the new Chinese HYVs, most of the irrigated land lay in the south. Wheat, although grown as a second crop, was primarily grown in the north China plain where only 20 per cent of it was irrigated. In the early 1960s, when rapid recovery from the fall in food output in 1959–61 was necessary, the Chinese government concentrated resources and new inputs in the areas of 'high and stable yields', which meant the irrigated southern rice areas where by 1977 80 per cent of the rice was sown with new varieties. In the 1960s and early 1970s little attention was paid to the north because of the lack of irrigation. But in the 1970s the great increase in tube wells led to the adoption of Chinese–Mexican wheats; by 1980 80 per cent of the wheat grown in the north China plain was irrigated.[26] Even so by the early

Table 10.15   High yielding variety rice and irrigation, 1975

	Rice irrigated (%)	Rice in HYVs (%)
Pakistan	100	40
Malaya	77	37
Sri Lanka	61	63
Indonesia	47	41
Philippines	41	68
India	40	36
Burma	17	7
Bangladesh	16	14
Nepal	16	18
Vietnam	15	n.d.
Thailand	11	11

HYV, high yielding variety.
Source: FAO, The State of Food and Agriculture 1977, Rome, 1978, pp. 2–16

Table 10.16   Modern varieties in Asia, 1982–3

	Rice		Wheat	
	Area in MVs (million hectares)	MVs as percentage of all rice	Area in MVs (million hectares)	MVs as a percentage of all wheat
Communist Asia	33.4	81.0	8.9	30.6
Non-Communist Asia	36.4	44.9	25.4	79.2
Near East	0.1	8.4	7.6	30.6
All Asia	69.9	56.8	41.9	48.7

MV, modern variety.
Source: M. Lipton and R. Longhurst, New Seeds and Poor People, London, 1989, p. 2

1980s a relatively small proportion of China's wheat was sown with improved varieties (table 10.16).

Although the new high yielding wheats and rices have received most attention, there have been improvements in the other food crops grown in Asia. By the early 1980s one-third of China's maize was sown with hybrid varieties, one-third of the rest of the continent. In India nearly half the area in sorghum and pearl millet was in HYVs.[27]

Although the distribution of irrigation seems to have been the major determinant of the rate of adoption by region, there are certainly other factors influencing differences between areas and farmers. The adoption of the HYV package of seeds, fertilizers and pesticides costs a great deal more per hectare than the use of traditional varieties – in Bangladesh, for example, 60 per cent more. It was thought in the early years of the introduction of the new varieties that large farmers, with greater access to credit and more able to run risks, were adopting the HYVs but not small farmers. But as data from the mid 1970s has become available it is clear that occupiers of small holdings have adopted the new seeds, although often with a time lag. Nor has the HYV rice always been confined to the irrigated areas. In the Philippines in 1975 78 per cent of irrigated rice was sown with IRRI rice; but so was 50 per cent of the rain-fed rice. Of considerable importance in determining the rate of adoption has been the rice–fertilizer price ratio. There are considerable variations in the price of rice and also the local price of fertilizer, owing either to market forces or to state intervention. The more favourable the ratio, the more likely it is that the new seeds and fertilizer will be adopted. Thus in Burma there was little incentive to increase costs by using new seeds and fertilizers, for the price of rice remained fixed by the government from 1952 to 1975, and farmers found that even the low cost traditional rices made a loss. In contrast in Pakistan fertilizer is cheap relative to rice prices.[28]

THE USE OF FERTILIZERS AND MANURE

In 1950 the supply of plant nutrients to crops was provided in a number of ways. Wet rice receives nitrogen from blue-green algae; it is estimated that these free living bacteria can fix up to 78 kilograms of nitrogen per hectare. In parts of East Asia a green crop was ploughed in, and in China soyabeans and elsewhere beans and peas, had bacteria associated with nodules on their roots that added nitrogen. In the lower rivers and deltas flooding added some nutrients in suspension and solution. Chemical fertilizers were not of any importance except in Japan, Korea and Taiwan. Elsewhere animal manure was the principal source of plant nutrients, but the supply was almost certainly low except in China. In China pig dung and human excreta were used as organic manures in the 1950s, as they had been of course before 1950. In 1957 chemical fertilizers were still hardly used in China, providing only 4 kilograms per hectare of arable; but organic manures provided between 36 and 110 kilograms of plant nutrients, a level far exceeding that obtained from this source elsewhere in Asia. Indeed the amount of

Table 10.17   Chemical fertilizers in Asia, 1950–88 (kilograms per hectare of cropland)

	1950	1961–5	1972	1979	1988
Japan	146.2	305.2	389.5	477.1	415.1
South Korea	52.5	157.0	288.9	383.6	399.9
North Korea	–	78.3	176.2	336.0	338.1
Taiwan	87.8	201.8	150.3	261.1	n.d.
China	4.0	13.3	45.5	109.2	262.1
Malaysia	0.4	9.4	35.5	103.2	150.8
Pakistan ⎫	0.2	3.4	22.8	51.9	83.2
Bangladesh ⎭		4.4	20.0	44.6	86.2
Indonesia	1.0	8.4	28.9	44.1	112.8
Philippines	0.5	13.3	25.6	34.6	63.3
India	0.5	3.7	16.7	29.6	65.2
Thailand	0.3	2.2	10.8	17.4	38.6
Burma	0.02	0.7	4.6	10.5	n.d.
Asian average[a]	1.6	11.8	31.0	61.5	n.d.
World average	12.4	27.9	54.3	77.1	98.7

n.d., no data.
[a] Excluding China and Japan.
Sources: FAO, *Production Yearbook 1957*, vol. 11, Rome, 1958; *Production Yearbook 1962*, vol. 16, Rome, 1963; *The State of Food and Agriculture 1970*, Rome, 1971; R. C. Hsu, *Food for One Billion: China's Agriculture since 1949*, Boulder, Colo., 1982, p. 57; FAO, *Fertilizer Yearbook 1989*, vol. 39, Rome, 1990

organic manure produced doubled between 1952 and 1977, largely through an increase in the number of pigs kept, many of them on private plots; organic manures still provided two-thirds of all the plant nutrients. But this was not without its cost. The acute lack of fodder meant that in 1957 about 20 per cent of China's unprocessed grain was being used for feeding livestock. Further, the collection of pig manure, its fermentation and its distribution to the fields takes a remarkable amount of labour, estimated at between 20 to 40 per cent of the working year.[29]

Between 1950 and 1965 there was a considerable increase in the consumption of chemical fertilizers in most countries in Asia, but the level was still low in 1965 except in Japan, Korea and Taiwan (table 10.17). In the Philippines, for example, chemical fertilizers were used on only 5 per cent of the cultivated area before 1965, and in Thailand in 1963 only 9 per cent of farmers used any chemicals; indeed only one-third applied even farmyard manure. Since the early 1960s, with the spread of HYVs and their need for heavy fertilizer applications, the consumption of fertilizer has risen dramatically in nearly all the Asian countries. The average rose more than fivefold between 1961–5 and 1979 (table 10.17) and at a prodigious rate in some countries which had low levels; in Pakistan, for example, the level rose by a factor of 15. But there was still a marked gap between East Asia and the rest of the continent. In the 1980s, however, there has been little increase in applications in East Asia, but they have continued to increase elsewhere in Asia.[30]

Initially most Asian countries relied upon imports of fertilizer; but India now produces half her needs. In China the 1960s saw the rapid growth of fertilizer consumption, which was initially either imported or made in small factories in the communes. In 1972 the Chinese government purchased 13 urea plants from abroad. Because the heavy application of organic manures provides enough potassium and phosphorus – half all plant nutrients came from animals – most of the chemical fertilizer applied is straight nitrogen. There are, as noted, still very wide variations in Asia in the use of fertilizer per hectare (table 10.17). Few HYVs receive the recommended dosage, and it is often of reduced value because it is applied at the wrong time. Furthermore many farmers have found that greater financial returns are obtained by using suboptimal applications.[31]

IRRIGATION

Irrigation has been part of Asian agriculture for several millenia, and a much higher proportion of the arable land (32 per cent) is irrigated than in Africa or Latin America. There are also very marked differences within Asia in the amount of land irrigated. Pakistan has the highest proportion but with this exception the ratio roughly reflects population density. The Republic of Korea, Japan and Taiwan have more than half their arable irrigated, Kampuchea, Burma, the Philippines, Laos and Malaysia have less than one-fifth (table 10.2). It should be remembered that over two-thirds of Asia's arable is not irrigated and that much of the irrigation is recent; in Pakistan and the Indian Punjab it was begun by the British at the beginning of this century.

The function of most irrigation systems in Asia is to secure the water supply for the main crop during the monsoon because of the unreliability of that rainfall. Storage played a small part in traditional Asian irrigation systems; their main function was to impound river flow during the period of high flow and divert river water by canals to the farmers' fields. In the last forty years more emphasis has been put upon storage reservoirs and multi-purpose water control. Thus the Chinese have since 1949 been attempting to control the floods of the Yellow River by a series of dams in its upper reaches, to produce hydroelectricity, reduce silting and salinity and provide irrigation water. A principal problem of irrigation systems is their inefficiency. In India 60–70 per cent of the water entering the system is lost through evaporation in storage and transit and by seepage through the canal bottoms; a figure of 50 per cent has been cited for China. Indeed the 'high and stable yielding' areas of China – those with both efficient irrigation and underdrainage – cover only 22 million hectares, about one-half of the area irrigated. Without some means of removing irrigation water land can easily become waterlogged and salts are deposited on the surface, initially reducing yields and eventually causing the land to be abandoned. Substantial areas in Pakistan were suffering from salinization in the 1960s, although the spread of tube wells since has reduced the problem. Salinity is a major problem in north China; in addition the Yellow River runs through the loess region in its upper reaches, and silting in the lower reaches clogs irrigation canals.[32]

The advantages of irrigation are considerable. First it helps reduce the variations in yield caused by fluctuations in rainfall. Second, efficient irrigation systems allow a second crop. Third, a controlled supply of water is necessary for the high yields of the new varieties, for without adequate moisture the value of fertilizers is much reduced. Fourth, irrigation allows crop growth in areas formerly too dry. The benefits of irrigation are clear. In Asia (excluding China) only one-third of the rice area is irrigated, but produces 60 per cent of the output. It is not surprising that in both India and China government policy has directed new inputs – seed, fertilizer and machinery – to these areas, and great attempts have been made to extend the irrigated area.[33]

It is difficult to trace accurately the expansion of the irrigated area. Some data record the arable area irrigated, others the gross sown area, thus including the multiple cropping that data for the net sown area exclude. There is also a distinction to be made between the area for which irrigation water can be made available and that on which it is actually used. The extension of the area in China and India is generally agreed to have been considerable (table 10.18). In China in the 1950s and 1960s the communes organized labour in the slack season to undertake not only local water control works but also large-scale works on the Yellow River. In both countries there has been a significant change in the means of obtaining water. In India and Pakistan canals have been the major means of distributing water from major rivers. In the Deccan, and in parts of Tamil Nadu and Sri Lanka, dams across small streams were important. In the last forty years tube wells have greatly increased

Table 10.18 Irrigation in Asia, 1950–89

	1950		1988–9	
	Area (million hectares)	Percentage of arable	Area (million hectares)	Percentage of arable
China	26.1	26.7	44	46.2
India	20.8	18.0	41.8	24.9
Pakistan	8.9	36.4	15.6	77.7
Bangladesh			2.2	23.9
Indonesia	5.0	28.2	7.5	34.9
Korea, South	0.57	28.5	1.3	57.1
Thailand	0.88	11.4	4.0	20.0
Malaya	0.11	5.0	0.34	13.3

Sources: R. C. Hsu, *Food for One Billion: China's Agriculture since 1949*, Boulder, Colo., 1982, p. 72; E. Dayal, 'Impact of irrigation expansion on multiple croping in India', *Tijdschrift voor Economische en Sociale Geografie*, **68**, 1977, pp. 100–9; FAO, *Production Yearbook 1957*, vol. 11, Rome, 1958; FAO, *Production Yearbook 1989*, vol. 43, Rome, 1990

Table 10.19 Irrigation in Pakistan and India

	Canals	Tanks	Tube wells	Total
Source of irrigation water, Pakistan (thousand cubic metres)				
1965–6	68		12	80
1976–7	72		36	107
India, area commanded by (million hectares)				
1950	8.2	3.6	5.9	
1975–6	13.7	3.9	14.3	

Source: B. H. Farmer, *An Introduction to South Asia*, London, 1984, p. 182; Mahmood Hasan Khan, *Underdevelopment and Agrarian Structure*, Boulder, Colo., 1981, p. 43

in numbers, tapping groundwater and also acting as a means of reducing waterlogging. They have spread at a prodigious rate: in Tamil Nadu the number of tube wells increased from 14,000 in 1951 to 743,000 in 1976, in north China 1 million were dug in ten years and in Pakistan and India they have become of major importance (table 10.19).[34] Water in wells can of course be lifted by human or animal labour, but in

Table 10.20 Changes in the intensity of cropping in Asia, 1950–80

	c.1950	c.1980
India	110	118
Bangladesh	134	141
Pakistan	111	121
Nepal	125	117
Malaya[a]	101	160
China	130	150
Taiwan	151	180[b]
Burma	107	111[c]
Philippines	126	136[d]
Thailand	–	101[e]

[a] Rice only.
[b] 1956–60.
[c] 1965–6.
[d] 1960.
[e] 1966.
Sources: B. L. C. Johnson, *Development in South Asia*, Harmondsworth, 1983, p. 62; D. S. Gibbons, R. de Koninck and Ibrahim Hasan, *Agricultural Modernization, Poverty and Inequality: The Distributional Impact of the Green Revolution in Regions of Malaysia and Indonesia*, London, 1980, p. 6; D. Dalrymple, *Survey of Multiple Cropping in Less Developed Nations*, United States Department of Agriculture, Washington, D.C., 1971

China, India and Pakistan wells have been increasingly powered by diesel or electricity. In India some 45 per cent of villages now have electricity, in China half the production teams.[35]

The expansion of irrigation in Asia has had a major effect upon crop yields, helped the extension of double cropping and has been the centre of much of the rapid advance of the last forty years. It is likely that future expansion of food output will depend upon their increasing efficiency.

MULTIPLE CROPPING

One of the distinctive features of Asian agriculture is, and always has been, multiple cropping. But, as with other intensive features, it was best developed in the densely populated areas of East Asia and was, and is, less common in the less densely populated areas. Thus the *multiple cropping index* – the ratio of the gross sown area to the arable area – was 150 in China in 1980, considerably higher in Taiwan, but 101 or less in Kampuchea and Thailand (see table 10.20).

The primary determinant of double cropping is temperature; but in Asia the length of the growing season only precludes the sowing of a second cereal crop in the more northerly parts of China, Korea and Japan. In most of Asia it is the seasonal distribution of rainfall that is limiting. It follows that irrigation would seem to be a major factor in the extent and expansion of double cropping. This needs some qualification, for much of the irrigation in Asia has only limited storage facilities and there may be insufficient water for a second crop. Nor may it always be possible to grow a second crop, for in some irrigated areas in Tamil Nadu cotton and sugar-cane are grown and occupy the land for most of the year. Even where irrigation has been provided specifically to introduce double cropping it may not take place. In the recently irrigated areas of north-east Thailand only 1.6 per cent of the land is used for a second crop because farmers have traditionally done other jobs in the dry season, and the income from these jobs is greater than that which can be obtained from growing a second crop.[36]

In many parts of Asia double cropping is not confined to the irrigated areas. A good summer monsoon may leave enough soil moisture for a second crop – although not generally for rice. Indeed in India 72 per cent of the land sown twice is not irrigated. However, the increase in double cropping since 1950 has generally been associated with the irrigated areas in India and elsewhere. The introduction of HYVs has increased the possibilities of double cropping, for the IRRI rice varieties mature more rapidly than most traditional varieties. The Chinese have preferred their indigenous high yielding rices to the IRRI varieties because they have an even shorter growing season, allowing not only double cropping but even triple cropping in parts of south China.[37]

Although 50 million hectares are double cropped in India, the cropping index has only advanced a few points since 1950, as is true in Bangladesh and Pakistan (table 10.20). The most spectacular increase has been in Malaya. Only a low proportion of Malaya's cropland is double cropped because oil palm and rubber occupy a large part of the farmland. In the 1950s Malaya relied upon rice imports to feed its population. Since then vigorous state schemes have greatly increased the proportion of rice cultivation that is irrigated, and associated with this has been a prodigious increase in the double cropping of rice. Combined with some increase in yields, Malaya's rice imports fell from 43 per cent of consumption in the 1950s to 10 per cent in the late 1970s.[38]

In China there has also been a substantial increase in double cropping. The cropping index has always been high in the south where a high proportion of the rice area was irrigated, and rice was followed by wheat or barley. Much less of the north China plain was irrigated, but since the late 1960s the spread of tube wells has increased the irrigated area and the double cropping index. In the south second crops have been further extended and triple cropping has been encouraged. In 1948 triple cropping was found only in Hai-nan island in the extreme south. In the late 1960s and early 1970s the Chinese government encouraged the spread of triple cropping, often into areas to which it was unsuited. Since 1976 triple cropping has declined in certain areas, such as Szechwan; peasants found that the net output was less than that obtained under double cropping. None the less cropping intensity has increased significantly in China, from 130 in 1950 to 150 in 1980; the index is well over 200 in parts of the south and this increase is thought to have accounted for 40 per cent of the increase in rice output between 1949 and 1975. It is to be doubted if the extension of multiple cropping has been as significant elsewhere, except in Malaya. However, the increase in the sown area has been of importance in raising food output in Asia, and is an important future source of increased output.[39]

LABOUR SUPPLY AND MECHANIZATION

Between 1950 and 1990 the agricultural population of Asia increased by 725 million, Africa by 216 million and Latin America by 31 million, the increase being respectively 68, 124 and 36 per cent. This growth (see table 4.6) has been a function of the rate of rural natural increase and the rate of migration to the towns. In Asia the rate of rural increase has been less than that in much of Latin America or Africa; but in few parts of Asia has the flight to the cities been so marked as in Latin America. Thus Asia's agricultural populations have increased more slowly than Africa's but more rapidly than Latin America's. The absolute numbers are formidable. The agricultural labour force in China in 1950 was about 170 million, in 1990 458 million; in India it was 102 million in 1951, 214 million in 1990. In a few countries the agricultural population has begun to decline; industrial growth has attracted rural labour in Taiwan, South Korea and Japan and numbers have also fallen in some countries in the Middle East.[40]

The growth of the agricultural population has of course presented agriculture with considerable problems, for population densities were already high in 1950 and the colonization of new land has not been sufficient to provide land for the increasing numbers. Thus it is not surprising to find that the numbers without land have increased substantially, although it should be noted that the definition of landless varies a great deal and the figures are not easy to interpret (table 10.21). Thus in Indonesia in the 1960s the number of very small holdings

Table 10.21   Percentage of rural households without land in Asia

	c.1954–5	c.1961–2	c.1971–2	c.1978	c.1983
India	30.8[a]	27.5	25.6	–	–
India	–	–	10.0	–	–
Pakistan	–	–	34.0	–	–
Bangladesh	14	17	–	28.8	–
Bangladesh	–	–	31	–	–
Bangladesh	–	–	17.5	–	–
Bangladesh	–	–	–	33.0	–
Bangladesh	–	–	–	11.07	–
Bangladesh	–	2	38	–	–
Indonesia	–	–	12.1	–	–
Thailand	–	–	–	4.7	10.0
China	10[a]	–	–	–	–
Taiwan	–	–	–	very low	–
Nepal	–	–	–	–	10.4
Sri Lanka	–	–	–	–	30.0

[a] 1948.
Sources: A. Bhaduri, 'A comparative study of land reform in South Asia', *Economic Bulletin for Asia and the Pacific*, **30**, 1979, pp. 1–13; M. Cain, 'Landlessness in India and Bangladesh: a critical review of national data sources', *Economic Development and Cultural Change*, **31**, 1983, pp. 149–69; B. L. C. Johnson, *Bangladesh*, London, 1975, p. 71; Iftikhar Ahmed, *Technological Change and Agrarian Structure: A Study of Bangladesh*, Geneva, 1981, p. 29; S. de Vylder, *Agriculture in Chains. Bangladesh: A Case Study in Contradictions and Constraints*, London, 1982; B. D. Mabry, 'Peasant economic behaviour in Thailand', *Journal of South East Asian Studies*, **10**, 1979, pp. 400–19; Asian Development Bank, *Rural Asia: Challenge and Opportunity*, New York, 1977, p. 55; 'Prospects for the economic development of Bangladesh in the 1980s', *Economic Bulletin for Asia and the Pacific*, 31, 1980, p. 102; R. Dernberger, 'Agriculture in Communist development strategy', in R. Barker and R. Sinha (eds), *The Chinese Agricultural Economy*, London, 1982, pp. 65–83; R. Montogmery and Toto Sugito, 'Changes in the structure of farms and farming in Indonesia between censuses, 1963–1973', *Journal of South East Asia Studies*, **11**, 1980, pp. 348–65; *Landlessness in Rural Asia*, Bhaka, 1987, pp. 15, 18, 24

increased substantially. Whether someone with 0.1 hecares is to be counted as landless or a farmer is a matter of debate. Consequently there are very different estimates of the extent of landlessness in Asia (note the various estimates for Bangladesh in table 10.21), although there is little doubt that in absolute terms it has increased since 1950 and that in some countries landlessness has increased faster than population growth (table 10.21).

One consequence of the growth of the agricultural population – both of farm families and wage labourers – has been that a significant proportion of the population is surplus to requirements on the farm. Thus in the late 1970s 35 per cent of the population of Bangladesh was unemployed or underemployed, and in Thailand in 1973 only 38 per cent of the available man-hours in agriculture were used. However, some of the underemployment so measured in Thailand reflects the lack of work carried out in the dry season. This is traditionally the period for work off the farm, and also for maintaining or improving irrigation works or clearing land for cultivation. In China the commune system allowed a massive mobilization of labour in the slack season. In 1957–8 some 100 million peasants were working in the winter on water control projects; in the 1970s this figure may have reached 150 million. The rapid expansion of tube wells in the north was due to this type of labour. However, since the introduction of the household responsibility system in 1979 and the abolition of the communes, investment in such public work has declined. Not only do individual farms not invest in such collective improvements, but state investment in agriculture has also fallen.[41]

But the utilization of labour in this way does not conceal the fact that China's rural population greatly exceeds the requirements of even the very labour intensive agriculture of that country. Chinese sources estimate that 10–20 per cent of rural labour was surplus in 1958 and about 30 per cent in the late 1970s. Indeed in the late 1970s Chinese authorities then feared that if the production responsibility system became established (a system which provides incentives at the household and individual level), some 50 per cent of the Chinese labour force in agriculture would be superfluous; since 1979 up to 20 per cent of Chinese peasants have left agriculture to find jobs in rural industry; but it still estimated that 30 per cent of the labour force is surplus to requirements.[42]

One consequence of the increase in the labour force in Asia, and particularly the growth of landlessness, has been a fall or stagnation in real wages in agriculture; this occurred in most Asian states in the 1970s, except in Taiwan and Korea where the labour force was in decline. Under these circumstances there seems little sense in pressing policies of mechanization in Asia, for this may further compound the problem. In northern China a farm was equipped for experimental purposes with American farm machinery, herbicides and pesticides; what had been done by 300 peasants was then done by twenty workers. In the very different conditions of central Thailand one man with one tractor can cultivate 20–24 hectares; without the tractor it would need a buffalo and 20–30 men. Even very simple changes can have disastrous consequences. In Java the landless population depend upon harvesting for much of their income. When the ani-ani (a hand knife) is used, harvesting rice needs 150 men per hectare; when the sickle is used, only fifty-five.[43]

There are of course cases for the use of machinery. Where double cropping is introduced it may be essential to use a tractor, for the time between harvesting and sowing the second crop is so short; similarly, some very heavy soils can only be cultivated mechanically. In parts of Malaya, which has not mechanized even though there have been rapid increases in irrigation, double cropping and the use of chemical fertilizers, the introduction of a second rice crop has prevented the use of straw as a bullock feed and encouraged some farmers to use tractors. There is also a case for the use of machinery in the few sparsely populated areas, such as Manchuria.[44]

But the decision to adopt tractors and other implements is not taken at the national level, except in China. It is thus on larger farms, particularly in the Indian and Pakistan Punjab, that the use of machinery has proceeded most rapidly, and indeed where the number of tractors was increasing before the introduction of new HYVs. Thus in India there has been a dramatic increase in the number of tractors in use, from 8500 in 1951 to 418,000 in 1981 and 925,000 in 1989. But the impact of this has been limited. In 1974 half the tractors in India were in Punjab, Haryana and western Uttar Pradesh, and of the 141 million hectares cultivated in India in that year, only 2 million were ploughed with tractors. Indeed in most of Asia the tractor provides only 4 per cent of the energy used on the farm; only in South-west Asia does the proportion approach that of Latin America (table 8.2).[45]

In China little attention was paid to mechanization in the 1950s, although at times the Soviet model of collectivization and large-scale mechanization was urged. Since the mid 1960s, however, the use of machinery has increased. China appears to have more tractors per hectare of cultivated land than either Africa or Latin America (table 10.22). Nor do these figures include the 'walking tractors', an adaptation of the Japanese rotary tiller, which have greatly increased in number since 1965.[46] By 1981 it was estimated that 46 per cent of China's sown area was tractor ploughed at least once (compared with 1.6 per cent in 1960). A mechanical rice transplanter was introduced in 1958 but in the late 1970s was used on only 0.7 per cent of the rice area, and only 2.6 per cent of the grain area was harvested by machine, mainly on the state farms of the north east. Less emphasis has been placed upon mechanization since 1978 because of the fear of unemployment. Indeed visitors have noted that tractors are used more for transport than in the fields.[47]

Table 10.22 Use of tractors and harvesters, major regions (per 1000 hectares of arable)

	Tractors		Harvesters	
	1961–5	1988	1961–5	1988
Africa	1.2	3.0	0.12	0.28
Latin America	3.2	7.7	0.5	0.79
China	0.6	9.1	–	0.36
Rest of Asia	0.45	11.9	0.05	3.7

Sources: FAO, *Production Yearbook 1989*, vol. 43, Rome, 1990; *Production Yearbook 1976*, vol. 29, Rome, 1977

Thus although much attention has been paid to the progress of mechanization in Asia, and particularly to its adverse consequences in Pakistan and the Indian Punjab, as yet only a small part of Asia's land is farmed with machines (table 8.2). However, the Food and Agriculture Organization estimates suggest that more tractors are used per hectare than in Africa or, surprisingly, Latin America (table 10.22).

CONCLUSIONS

The increased output of Asian agriculture has matched that of Latin America since 1950, and in all but a few countries has kept up with population growth, though a significant surplus has been obtained in only a few countries. But the increase in output has been obtained in a very different way to that of Latin America, or the more modest increases in Africa. Colonization of new land has been important in only a few areas, and the reduction of fallow has been confined to a few countries. Instead Asian farming has increased output by traditional means – more irrigation, more intensive cultivation, more multiple cropping – and by the use of new technologies, improved seed and chemical fertilizer. For all the growth, neither poverty nor hunger have been banished. It is likely that in the future Asia's major problem will be to absorb its growing numbers rather than simply to raise food output, although that of course will continue to be necessary.

11
Trade and Aid

Although food production has grown more rapidly than population in the world as a whole and in all the major regions except Africa, in many individual countries food production per caput has fallen in the last twenty years. Yet in all but a few countries available national supplies per caput are above the level of the 1950s. This is of course because home production has been supplemented by imports of food; this trade is now considered.

THE DEVELOPMENT OF TRADE IN FOODSTUFFS

Until the middle of the nineteenth century the high cost of transport and the impossibility of preserving many foods for long periods meant that there was little international trade in foodstuffs. In Western Europe before the eighteenth century only 1 per cent of cereal output crossed national boundaries. That proportion is now higher, but the great bulk of all foodstuffs are still consumed in the country of origin; in 1983–5 only 12 per cent of food output entered international trade. In the middle of the nineteenth century there was little long distance movement of foodstuffs (figure 11.1) and much of this was from Eastern to Western Europe, where the Low Countries had long imported grain and Britain had been a net importer since the late eighteenth century.[1]

After 1850 there was an increase in the trade in foodstuffs (figure 11.1). The export of people and capital from Europe to North America, Australasia and South America, the rising populations and incomes of Western Europe and the abundant land of the European settlements overseas combined to produce a flow of cheap cereals, meat and dairy products to Western Europe and in particular to Britain. This was made possible by marked falls in oceanic and overland freight rates in

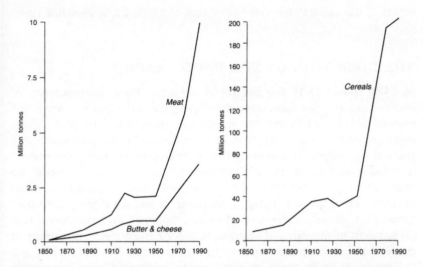

Figure 11.1 World exports of meat, butter, cheese and temperate cereals (wheat, maize, barley, oats, rye) *Sources*: R. M. Stern, 'A century of food exports', *Kyklos*, 13, 1960, pp. 44–64; FAO, *Trade Yearbooks*, Rome, various dates

the late nineteenth century and the introduction of refrigeration, which allowed frozen meat and dairy products to be moved. The export of cereals, which in volume greatly outweighed that in other foodstuffs (figure 11.1), was dominated initially by the United States, Russia and Eastern Europe (the Danube countries), but later they were joined by Canada, Argentina and Australia. India was also a major exporter, with 10 per cent of world exports of grain in the 1880s. The trade in grain increased ninefold between the 1850s and the eve of the First World War.[2] Between the two world wars the trade in foodstuffs stagnated or even declined (figure 11.1). However, notable gains were made by some products, such as coffee. Most of the trade in meat, dairy products and cereals moved from the temperate regions of European settlements overseas to temperate Western Europe. But in the later nineteenth century there had been a substantial increase in the flow of tropical and subtropical foods to Western Europe and the United States. These included tea, coffee, cocoa and sugar; bananas and other fruits increased in the very late nineteenth century.

Since the end of the Second World War the trade in foodstuffs has increased at an unprecedented rate, and much faster than food output or population. In the 1950s world exports of temperate grains (figure 11.1) were little above the level of 1909–13, between the early 1950s and the late 1980s the volume has risen sixfold and the meat trade, which in the early 1950s was below that of 1909-13, is now five times that level.[3]

THE DIRECTION OF THE FOOD TRADE

In 1988 (table 11.1) the developed countries took three-quarters of all food imports (by value), the developing countries but a quarter; the pattern established in the late nineteenth century still survives. Their most important single import is meat, followed by dairy products and tropical beverages. The imports of the developing countries are smaller in volume and value and different in composition. Cereals make up nearly three-quarters of the volume of imports, over one-third the value; oilseeds and tropical beverages, products of the developing countries themselves, are not important. Meat and dairy products now constitute 30 per cent of the value of imports, reflecting the growing prosperity of parts of the developing world.

Cereals are by far the most important item in world agricultural trade – indeed they are only exceeded in value by petroleum. They need some comment. Wheat, maize, rice, millet, sorghum, barley, oats and rye are all consumed as food by man in some part of the world, but the coarse grains – those other than wheat, rye or rice – which enter international trade are used as livestock feed. Wheat and rice, however, although generally used as human foods can be used to feed livestock, although it is only in Eastern Europe and the USSR that a high proportion of wheat is used as feed.[4] Trade statistics do not reveal the use of the cereals. However, the proportion of all cereal imports that is coarse grains is generally inversely related to income. In 1980 the lower the income, the higher the percentage of imports that was in food grains (table 11.2). Thus the developing regions (except Latin America, the most prosperous) import mainly food grains, the developed regions feed grains. Of all grain consumed in the developed countries 60 per cent is fed to livestock, in the developing countries only 13 per cent.[5]

The volume of agricultural trade has greatly increased since 1950, and indeed has increased more rapidly than either population or food production. There have also been major changes in the direction of trade, particularly in that of cereals. In the 1930s Western Europe took nearly all the imports and the export trade was shared by North America, Eastern Europe, the Soviet Union, Australasia and temperate Latin America (table 11.3). But since 1950 Eastern Europe, the USSR. and Latin America have ceased to be net exporters of grain and have become importers. The export of cereals is dominated by North America and Australia which in the late 1980s provided nearly 60 per cent of all exports, the United States alone accounting for 38 per cent.

Table 11.1 The structure of food imports, 1988

	Developed countries				Developing countries			
	Weight (million tonnes)	Percentage	Value (US dollars million)	Percentage	Weight (million tonnes)	Percentage	Value (US dollars million)	Percentage
Cereals	112.1	39.0	18,500	12.6	115.9	71.9	18,158	37.4
Meat	10.6	3.7	29,859	20.3	3.1	1.9	5,137	10.6
Dairy and eggs	9.1	3.2	20,150	13.7	3.5	2.2	9,409	19.4
Fruit and vegetables	33.6	11.7	17,201	11.7	7.6	4.7	3,178	6.5
Sugar	14.5	5.1	8,431	5.7	13.6	8.5	3,562	7.3
Coffee, cocoa, tea	7.5	2.6	19,632	13.3	1.1	0.6	2,556	5.3
Oilseeds and cake	64.7	22.5	15,094	10.3	8.1	5.0	3,571	7.3
Wine and beer	7.1	2.5	8,827	6.0	0.6	0.4	638	1.3
Oilseeds	27.9	9.7	9,378	6.4	7.8	4.8	2,400	4.9
Total	287.1	100.0	136,273	100.0	161.3	100.0	48,609	100.0

Source: FAO, Trade Yearbook 1988, vol. 42, Rome, 1990

Table 11.2 Rice and wheat as a percentage of all cereal (volume), 1980

Africa	79.6	Latin America	51.2
Far East	76.4	Eastern Europe and USSR	45.8
Asian CPEs	73.5	Western Europe	32.5
Near East	71.0	Japan	23.9

CPEs, centrally planned economies.
Source: FAO, Trade Yearbook 1981, vol. 35, Rome, 1982

Table 11.3 World grain trade, 1934–88 (million tonnes: positive values net exports, negative values net imports)

	1934–8	1948–52	1960	1970	1979–80	1988
North America	5	23	39	56	127	119
Western Europe	–24	22	–25	30	–13	22
Eastern Europe and USSR	5	n/a	0	0	–40	–27
Australia and New Zealand	3	3	6	12	15	14
Latin America	9	1	0	4	–9	–11
Africa	1	0	–2	–5	–13	–28
Asia[a]	–2	–6	–17	–37	–63	–89

[a] Including Japan and China.
Sources: R. F. Hopkins and D. J. Puchala, 'Perspectives on the international relations of food', in R. F. Hopkins and D. J. Puchala (eds), The Global Political Economy of Food, Madison, Wis., 1978, p. 7; FAO, Trade Yearbook 1981, vol. 35, Rome, 1982; Trade Yearbook 1971, vol. 25, Rome, 1972; Trade Yearbook 1966, vol. 20, Rome, 1967; L. R. Brown, The changing world food prospect: the nineties and beyond, World Watch Paper 85, Washington, D.C., 1988

There have been equally striking changes in the major markets for cereals. Western Europe's imports – mainly of feed – rose until 1970 but the European Community (EC) policy of levies and high guaranteed prices for cereals has led since to a decline in imports and substantial if subsidized exports. Japan, however, has greatly increased her imports, owing partly to rising population but mainly to greater prosperity. Between 1960 and 1975 meat consumption per caput tripled, and this was only achieved by importing feed grains. Japan's self-sufficiency ratio fell from 90 to 73 per cent in this period.[6] More dramatic has been the emergence of the Soviet Union as an importer after a long history of exporting. In 1950 Soviet agricultural output was little higher than in 1913, but output has increased very rapidly since then. However, the former USSR has two problems. First, much of the grain acreage is east of the Urals, in very dry areas. There are thus marked year to year fluctuations in the harvest. Second, this output is insufficient, even in good years, to provide both bread and feed for livestock. Russian demand for meat has risen and grain imports have become necessary, beginning with large purchases from the United States in 1972 which reached 40 million tonnes in 1980. Although at a lower level in the 1980s (table 11.3) they remain substantial.

FOOD IMPORTS OF THE DEVELOPING COUNTRIES

Although the developed countries still take a greater part of world trade in agricultural products (table 11.1), a noticeable feature of the last thirty years has been the remarkable increase of food imports, particularly of cereals, into the developing countries. In the 1930s neither Asia – except Japan – Latin America nor Africa were net importers of grain (table 11.3), although of course some individual countries were. It should be noted, however, that consumption levels were lower than they are now. By the 1950s imports into Africa and Asia were growing slowly; they accelerated in the 1960s and 1970s.

This change in the direction of trade can be variously illustrated. Thus the developing countries – including China – took 1.5 per cent of world cereal imports in the 1930s, 20 per cent in the 1950s, 44 per cent in 1980 and one-half in 1988. The volume of these imports increased more than sixfold between the 1950s and the late 1970s. The developing countries' imports of wheat have increased rather more rapidly than the other grains; in the 1950s the developing countries accounted for 10 per cent of all wheat imports, in the 1980s two-thirds.[7]

The reason for these imports would appear to be simple. In much of Africa population growth has outpaced food production and imports have been necessary simply to maintain consumption levels. In other countries they have been needed to try and raise the very low per caput supplies. But this assumes that population growth alone accounts for the rise in food imports. This is far from being so, for in many developing countries incomes have risen substantially in the last twenty years; this is most noticeable in the oil exporters, but also in several countries in the Far East. This has led not only to increases in wheat and rice imports but also to cereal imports to feed livestock. Those developing countries which the World Bank describes as middle income not only accounted for 80 per cent of the increases in developing countries' cereal imports between 1960 and 1979, but also fed one-third of their domestic grain production to livestock and half their cereal imports.[8] In short cereal imports to the developing countries are not directly related to population growth or to need. This is further illustrated in figure 11.2, which shows net cereal imports per thousand of the total population in 1988. A few countries – very few – had a grain balance in surplus. These included, as might be expected, the traditional grain exporters of North America, Australia, Burma, Thailand and Argentina; but with EC export subsidies many European countries have become net exporters. Those developing countries with imports of 100 tonnes per thousand of the population or over include only the more prosperous countries – oil exporters such as Venezuela, Libya and Saudi Arabia or countries which have achieved some degree of industrialization such as South Korea.

SELF-SUFFICIENCY AND DEPENDENCY

In the early 1950s and early 1960s there was much discussion of the ability of the developing countries to increase food output and of the low level of national food supplies per caput; food imports were comparatively small and received little attention. The failure of the monsoon in India in 1964–6 and the steep rise in imports drew attention to the vulnerable situation of countries that relied upon imports. The great rise in grain prices in 1972–4 and the increase in imports led many to argue that developing countries should attempt to obtain self-sufficiency in food supplies.

Self-sufficiency is extremely difficult to measure;[9] if estimating self-sufficiency in food supplies is the aim, then data for all production and trade in all foodstuffs, converted into calorific values, is necessary and such data are rarely available. More commonly self-sufficiency ratios are calculated by comparing the value of production and domestic consumption of foodstuffs, as in table 11.4.

Using this method, consumption in the developing countries as a whole was met by internal production in 1983–5, and only in the Near East and North Africa was self-sufficiency not achieved. The fact that self-sufficiency was attained did not of course mean that nutritional needs were met. Nor was the Near East and North Africa's failure to achieve self-sufficiency of much significance, for it reflects the import of livestock products and livestock feeds which an increasingly affluent population demands.

Between the early 1960s and the mid 1980s the developing countries as a whole experienced a small decline in the self-sufficiency ratio. The

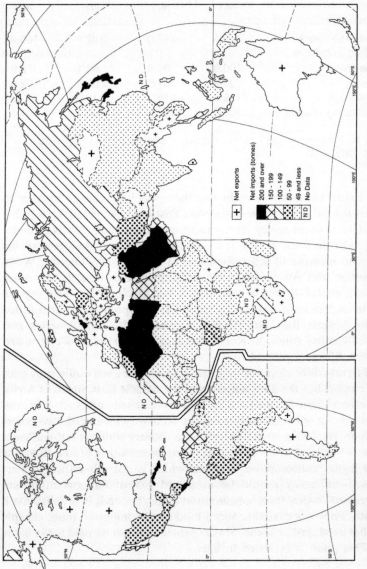

Figure 11.2 Net grain imports, 1988; tonnes per thousand of the total population *Source:* FAO, *Trade Yearbook 1988*, vol. 42, Rome, 1990

Table 11.4 Self-sufficiency ratios in developing countries

	1961–3 (%)	1969–71 (%)	1979–81 (%)	1983–5 (%)
Sub-Saharan Africa	119.8	117.0	102.9	100.8
Near East and North Africa	100.9	97.4	80.1	75.6
Asia	100.9	101.9	99.5	102.0
Latin America	119.9	115.7	113.0	113.9
All developing countries	106.6	105.7	100.3	101.1

Source: N. Alexandratos (ed.), *World Agriculture: toward 2000, an FAO study*, London, 1988, p. 33

fall in Latin America was slight and there was little change in Asia; in contrast the fall in Africa and the Near East has been precipitous (table 11.4). Cereal crops are by far the most important contributors to home food production and to the trade in foodstuffs, and thus provide an alternative measure of self-sufficiency. In 1988 North America, Western Europe and Australasia were self-sufficient in cereals, the other major regions all having a ratio below 100 (table 11.5). The regions most dependent upon imports included the oil exporters of North Africa and the Middle East, the Andean republics of South America and much of Central America, Korea and Japan and parts of Western and Southern Africa (figure 11.3).

It is clearly difficult if not impossible to measure self-sufficiency with any accuracy; but the data suggest that the Middle East, much of Africa and western Latin America are the regions least self-sufficient. But even if the degree of self-sufficiency could be accurately measured, two points remain at issue. First, it is debatable if a country which is self-sufficient but has a low level of consumption is less precariously placed than one with a higher consumption level but which is not self-sufficient. Perhaps self-sufficiency should be measured against national minimum requirements rather than consumption alone. Second, it is debatable if self-sufficiency is a desirable aim if food can be imported more cheaply than that produced at home, always provided it can be paid for in some way. This point is returned to later.

FOOD AID AND FOOD STOCKS

The rising price of food grains has been one reason why many developing countries have sought to maintain self-sufficiency. But not all

Table 11.5 Grain self-sufficiency, 1988

	Home production (million tonnes)	Exports (million tonnes)	Imports (million tonnes)	Consumption (million tonnes)	Home production as a percentage of consumption
Developing countries					
Africa	63.1	0.7	17.5	79.9	79
Near East	73.5	5.7	31.0	98.8	74
Far East	349.2	8.6	24.7	365.3	96
Asia CPEs	384.5	5.5	22.5	401.5	96
Latin America	108.3	10.6	19.7	117.4	92
Other developing countries	0.04	–	0.4	0.4	10
Total, developing countries	978.64	31.1	115.8	1063.3	92
Developed countries					
USSR and East Europe	277.7	5.5	43.7	315.9	88
North America	242.4	123.9	2.4	120.9	200
Western Europe	195.7	54.9	35.8	176.6	110
Australasia	23.0	14.9	0.1	8.2	280
Other developed countries	25.4	1.5	30.0	53.9	47
Total, developed countries	764.2	200.7	112.0	675.5	113

CPEs, centrally planned economies.
[a] USSR only.
[b] USA only.
[c] Japan only.
[d] Japan, South Africa, Isreal.
Sources: FAO, *Production Yearbook 1989*, vol. 43, Rome, 1990; FAO, *Trade Yearbook 1988*, vol. 42, Rome, 1990

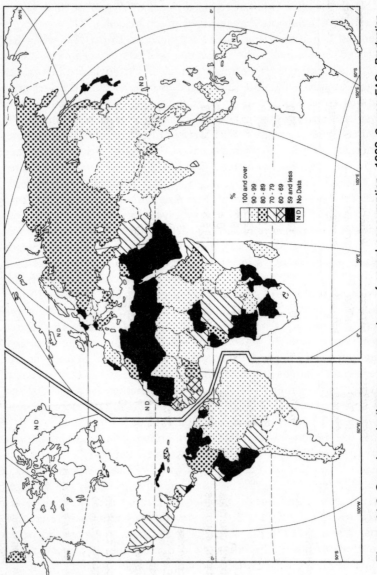

Figure 11.3 Cereal production as a percentage of cereal consumption, 1988 *Source:* FAO, *Production Yearbook 1989*, vol. 43, Rome, 1990; *Trade Yearbook 1988*, vol. 42, Rome, 1990

%
100 and over
90 - 99
80 - 89
70 - 79
60 - 69
59 and less
No Data

the cereal imports of the developing countries have to be paid for at the market price. Since 1954 substantial amounts of food – mainly cereals and milk products – have been sent to developing countries, primarily from North America either at reduced prices or as grants. Such apparently generous behaviour has received much criticism.[10]

Food aid began after the end of the Second World War when half the United States economic assistance to Europe's recovery was in the form of food. It became more formal in 1954 when a law passed through Congress – PL 480 – permitted the sale of United States grain on various concessionary terms to friendly and needy countries. This law was one response to the huge surpluses of grain accumulating in the United States, surpluses which were in turn a result of United States government policies first introduced in 1933 and variously modified since. In the late 1920s and the 1930s world recession reduced United States farmers' exports and led to very low prices and much poverty. In 1933 the United States government introduced legislation that paid farmers to withdraw land from cultivation and grow legumes or adopt soil conservation policies, and in the case of cotton introduced an area quota. But the government was also prepared to buy all cereals which could not be sold above a given price on the open market, and stored these surpluses. In the Second World War and its immediate aftermath there was no problem in disposing of surpluses. The adoption of new technologies led to great increases in output in the 1940s, but the rapid recovery of European agriculture reduced the overseas markets for United States exporters, and PL 480 was one means of reducing the surpluses. Throughout the 1950s and 1960s stocks in the United States – and to a lesser extent Canada – were high, and acted as a world reserve that maintained price stability in the world cereal market. In the 1950s and 1960s approximately half all cereal imports into the developing countries were in the form of food aid (figure 11.4).[11]

In the late 1960s, however, successive United States administrations attempted to reduce the burden of agricultural support, and land was withdrawn from cultivation. In 1972 and 1973 this, combined with unexpectedly large Russian purchases of United States grain and droughts in parts of Africa and South Asia, led to a rapid rise in cereal prices and, because of the shortage in many developing countries, to increased imports. When by 1976 the period of high prices was over and world food output up again, the imports did not decline, because demand continued from the better off developing countries, particularly the oil exporters, who had benefited from their raising of oil prices. After the devaluation of the American dollar in 1971 and the increased cost of United States oil imports, the United States government were determined to compensate by promoting United States food exports. For all

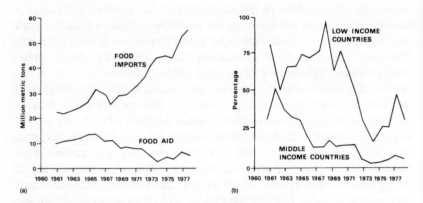

Figure 11.4 (a) Imports of food and food aid; (b) food aid as a percentage of food imports *Source*: World Bank, *World Development Report 1981*, Washington, D.C., 1981

these reasons food aid as a proportion of all developing countries' food imports has fallen. It was 57 per cent in 1961–2, 38 per cent in 1970 and 18 per cent in 1977; by 1982 food aid was only 5 per cent of the international trade in cereals.[12] However, food aid to the low income countries remains a high if declining proportion of their total food imports (figure 11.4).

In the 1950s and 1960s the United States was the major supplier of food aid. In 1965, for example, it provided 94 per cent of all food aid. But from 1968 EC policies of price support were also creating surpluses, particularly of dairy goods; the EC began to provide aid, notably of milk products but by the late 1970s of cereals as well, so that by 1976 the United States share of all food aid had fallen to 68 per cent. During the 1980s EC food aid increased and by the late 1980s the United States provided only half the value of all food aid, the EC 28 per cent. Together with Canada they accounted for nine-tenths of all food aid.[13]

Over the last thirty years more than ninety countries have received some form of food aid; between 1945 and 1980 $30 billion of aid has been distributed. There has been much criticism of the direction and consequences of aid. In the 1950s and early 1960s most aid was in the form of bilateral arrangements between the United States and recipient countries. It has been argued that much of this aid did not go to those countries most in need, but instead to those with which the United States had close military or political relationships, or to those countries which already traded with the United States.[14]

Since the mid 1960s an increasing proportion of food aid has gone through the agency of the World Food Programme, run by the Food and Agriculture Organization (FAO), which receives food from donor countries but decides its destination. Since 1975 the United States Congress has required that 70 per cent of United States food aid should go to the poorer states – in 1981 those states with a gross domestic product per caput below $795. These countries received four-fifths of all food aid in 1981. But there have been numerous other criticisms of food aid. It has been argued that imports of concessionary grains have in some cases reduced the price of local products and hence been a disincentive to the increase of domestic food output. The fall in the wheat area in Columbia in the 1960s is a much quoted instance of this, although a study of India, a major recipient of food aid, suggests that adverse effects on prices only operated over very short periods.[15] Approximately 70 per cent of food aid is sold at market prices to importing governments and the proceeds used to finance development projects. The remainder, described as project food aid, is supplied free and distributed in various ways – as free school meals, in health centres for mothers and young children and as food for labourers on land improvement schemes. This aid does not always benefit the recipients. Thus supplementary feeding to children and infants is balanced by their receiving less food at home. Some foods sent as aid are inappropriate – such as Swiss cheese to Biafra. Further, the cost of transporting aid to the developing countries uses up a formidable proportion of total aid. In 1985 the ocean freight upon grain sent to East Africa was 30 per cent of the purchase price in the United States, and 50 per cent where there was no bulk handling in the receiving port. Current opinion appears to be that, except for emergency relief, food aid would be more beneficial in terms of untied financial grants or loans.[16]

In the 1950s and 1960s the price of wheat and other cereals was rising but did not fluctuate greatly from year to year, for the United States reserves could be fed into the system in times of shortage. With the rundown of these stocks and the growing proportion of commercial exports in total trade, there has been a renewal of a need for a central world food agency to maintain reserves. Although the FAO now supports a small emergency fund there has been no success in forming a world food reserve because of the conflicting interests of the developed exporters and the developing importers. In the 1980s carry over grain stocks – the amount in store before the new harvest – reached very high levels (table 11.6). Most of these reserves are held in developed countries, in 1987–9 two-thirds; and the United States alone, in spite of its efforts to curtail output, held 39 per cent of the total.[17]

Table 11.6 Carry over cereal stocks

Period	Million tonnes	Percentage of world consumption	Period	Million tonnes	Percentage of world consumption	Period	Million tonnes	Percentage of world consumption
1961	185	22	1976-7	166	18	1986	420.5	24
1965	159	17	1977-8	177	19	1987	450.9	26
1969-70	179	23	1978-9	200	21	1988	397.5	23
1970-1	144	17	1980	256	18	1989	288.1	16
1971-2	168	19	1981	252	16			
1972-3	119	14	1982	275	18			
1973-4	100	13	1983	342	20			
1974-5	108	12	1984	281.9	16			
1975-6	124	14	1985	335.5	19			

Sources: FAO, The State of Food and Agriculture 1978, Rome, 1979, pp. 1–17; The State of Food and Agriculture 1982, Rome, 1983; The State of Food and Agriculture 1975, Rome, 1976, p. 74; The State of Food and Agriculture 1989, Rome, 1989, p. 13

AGRICULTURAL EXPORTS FROM THE
DEVELOPING COUNTRIES

In 1950 agricultural produce accounted for much of the export earnings of the developing countries. This proportion has greatly declined
with the rise of oil exports and also, from a number of countries, of
manufactured goods exports; thus agricultural exports accounted for 50.
per cent of all export earnings in 1965, 23 per cent in 1980 and currently only one-tenth. None the less agricultural products still account
for 20 per cent of all Africa's exports, 10 per cent of the Far East's and
28 per cent of Latin America's, and in many countries the dependence
is far greater (figure 11.5). But the agricultural exports of the developing
countries, although increasing substantially in volume and value, have
declined as a proportion of world agricultural trade. In 1948–52 the
developing countries accounted for about 53 per cent of all agricultural
exports, in 1961–3 44 per cent (table 11.7) and by 1989 only 27 per
cent. One ominous consequence of this, combined with the steep rise
in food imports, has been the decline of the developing countries' ability
to pay for food imports with agricultural exports. Indeed by 1982 the
developing countries became net importers of agricultural products and
only Latin America had a sizeable surplus. Of course other products
are exported, but the dependence of many developing countries upon
agricultural exports makes it increasingly difficult to finance food
imports[18] (table 11.8).

Many developing countries are vulnerable because of their dependence upon one or two products. Thus tropical beverages, which are
largely luxury items, made up 25 per cent of all developing countries'
agricultural exports in 1980 and 16 per cent in 1988; of eighty-seven
developing countries in 1980 half depended upon cocoa, coffee and tea
for at least 30 per cent of their agricultural earnings. But concentration
is not confined to these crops. North Korea derives over 70 per cent
of its agricultural exports from rice alone. In Bangladesh jute makes up
half of all agricultural exports and Senegal and Gambia still depend
upon groundnuts; over 90 per cent of Cuba's exports come from sugar.
Some countries have successfully diversified; thus Malaysia has added
oil palm to rubber – albeit at the expense of several West African states
– Thailand has reduced its dependence upon rice and Brazil has become the world's second soybean exporter. But it is the countries with
either minerals or manufactures to export which have been able to
finance food imports most easily. Hence in the Far East, where grain
imports are admittedly not a major item, Korea, Hong Kong and Singapore take nearly half all the cereal imports, and in Africa, Algeria,

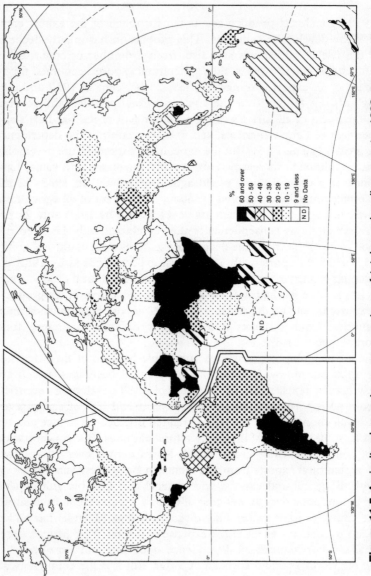

Figure 11.5 Agricultural exports as a percentage of total merchandise exports, 1988 *Source:* FAO, *Trade Yearbook 1988*, vol. 42, Rome, 1990

%
60 and over
50 - 59
40 - 49
30 - 39
20 - 29
10 - 19
9 and less
N D   No Data

Table 11.7 Share of world agriculture exports by value, 1961–3 to 1988

	1961–3 (%)	1972–4 (%)	1980 (%)	1988 (%)
Developed countries				
North America	19.7	23.1	21.4	20.7
Western Europe	17.6	26.1	36.2	42.5
Oceania	8.2	7.1	5.4	4.2
East Europe and USSR	7.9	6.0	4.2	3.8
Other developed countries	2.1	1.6	1.8	1.5
Total, developed countries	55.5	63.9	69.0	72.7
Developing countries				
Africa	9.1	6.5	4.5	2.8
Far East	12.5	8.3	8.5	9.9
Latin America	17.1	15.4	13.8	9.5
Near East	3.7	3.2	2.2	1.6
Asia CPEs	1.8	2.4	1.8	3.3
Other developing countries	0.3	0.3	0.2	0.2
Total, developing countries	44.5	36.1	31.0	27.3

CPEs, centrally planned economies.
Sources: FAO, *The State of Food and Agriculture 1975*, Rome, 1976, p. 68; FAO, *Trade Yearbook 1981*, vol. 39, Rome, 1982; *Trade Yearbook 1988*, vol. 42, Rome, 1990

Table 11.8 Developing countries: agricultural imports as a percentage of agricultural exports by value, 1970–2, 1980 and 1988

	1970–2	1980	1988
Africa	44	97	99
Far East	86	82	95
Latin America	32	44	36
Near East	97	363	339
Asian CPEs	105	215	100
All developing countries	60	95	92

CPEs, centrally planned economies.
Sources: FAO, *The State of Food and Agriculture 1982*, Rome, 1983, p. 19; *Trade Yearbook 1988*, vol. 42, 1988, Rome, 1990

Morocco and Nigeria a similar proportion; the Near East has dramatically increased its imports since the 1960s.[19]

The concentration upon relatively few exports is repeated when the major regions are considered. Some 40 per cent of Africa's agricultural exports are coffee and cocoa, 40 per cent of Latin America's are sugar, coffee and soybeans. The exports of the developing countries face a number of difficulties. First, those which are raw materials for manufacturing industry have been liable to be displaced by alternatives. Thus the rise of bulk carriers on land and sea has reduced the demand for jute sacks, and cotton, sisal and wool have been challenged by a variety of synthetic fibres. Although natural rubber remains a buoyant export, synthetic rubber manufacture has limited its rate of expansion. Second, much of the exports are commodities for which there is inelastic demand in the developed countries. During the economic expansion of the 1950s and 1960s in Europe and the United States, there was a continuing increase in demand for luxury commodities such as coffee, cocoa, bananas and tea, but by the early 1970s some argued that the market for these products was saturated. Third, output of many of the developing countries' agricultural exports fluctuate owing to variations in climate; thus prices oscillate violently. This is true of all agricultural products, but the impact is far greater when a country earns much of its export earnings from one or two products. Thus in the 1970s there have been numerous demands for the control of commodity prices. Fourth, about four-fifths of the agricultural products entering international trade are grown in both the developed and the developing countries. Only the products of the humid tropics are a monopoly of the developing countries. But developing countries have great difficulties in competing in North America and Europe, by far the most important markets for agricultural exports, because of the system of protection in the form of levies, subsidies, tariffs and guaranteed prices. The EC's protection has affected both developing countries and the traditional cereal exporters of North America and Australasia, both by imposing levies upon imports and by subsidising exports from the EC. Indeed it is the conflict between the EC and the United States which has prompted an as yet fruitless attempt to reduce, under the rules of the General Agreement on Tariffs and Trade, protection of agricultural produce. This, if achieved, would benefit both developed and developing countries, and particularly the consumer in the former region; in 1990 the subsidy to agriculture in the OECD countries was 44 per cent of the total value of output.[20]

For these and other reasons the terms of trade – the price of the developing countries' exports divided by the price of the manufactures they import – have been unfavourable to the developing countries for much of the post-war period, notably in the late 1950s and much of the 1960s and 1980s, although they were strongly in their favour in the early 1950s and much of the 1970s.[21]

CONCLUSIONS

The period since the end of the Second World War has seen a great boom in agricultural trade and marked alterations in its direction. Since the early 1960s there has been a great increase in food imports by the developing countries. This has been due first to the rapid population growth and the inability of some countries, particularly in Africa, to maintain per caput food supplies from home production, and second to the imports of wheat, livestock products and livestock feeds into the richer developing countries. The latter has been possible only where there have been oil or manufactures exports. The poorer developing countries, often dependent upon one or two crops for much of their exports, have found it increasingly difficult to pay for food imports.

12
Conclusions

Undernutrition and malnutrition remain a problem; in the 1970s it was the Sahel that held the world's attention, in the 1980s Ethiopia and in the early 1990s Somalia. There continues to be widespread discussion as to how famine, undernutrition and malnutrition in Afro-Asia and Latin America could be reduced. Before considering events in these areas, it may be helpful to review what has happened in Europe, and particularly Western Europe.

Although poverty causes some undernutrition in Europe today, it is a very minor problem. Since the 1940s nutritional diseases have been those due not to deficiency, but to excess; clinical and nutritional investigations have dealt with the relationships between, for example, food intake and obesity, the hazards of excessive sugar consumption, the relationship between animal fats and heart disease, the need for dietary fibre and the benefits of the Mediterranean diet. Yet before 1800 food consumption levels in Western Europe appear to have been very similar to those now found in America and South Asia. At a national level few countries had available supplies above 2300 calories per caput per day. Cereals and roots provided over 70 per cent of all calories and livestock products a small proportion, albeit probably higher absolutely and proportionally than in much of tropical Africa and South Asia today. On the other hand famines – a shortage of food in specific areas that leads to an increase in the death rate – were by far less frequent than they had been in the seventeenth century and before. The end of the Napoleonic Wars saw food shortages throughout much of Western Europe; the last great famine was in Ireland in the 1840s, and famine then disappeared from European, although not from Russian, history; acute food shortages and serious loss of life occurred there in 1891–2, 1921 and 1932–4.[1]

The nineteenth century saw rapid population growth – although far short of the rates of increase found in the developing countries since

1950. However, available supplies per caput steadily increased and had exceeded 3000 calories before the end of the century in much of Western Europe. This increase was due to increased consumption of cereals and roots, but also of fruit and vegetables, oilseeds and sugar and especially of livestock products. In the later nineteenth century the *proportion* of all calories derived from livestock products began to increase, and that from plant foods decline. Since the early twentieth century there has been an *absolute* decline in the consumption of cereals and roots.[2]

The elimination of hunger in Western Europe has taken a remarkably long time. The first success was the reduction of famine. No doubt increases in agricultural productivity were partly responsible for this, but other factors were important. As transport improved in the seventeenth and eighteenth centuries food could be moved faster and more cheaply to areas of shortage. There was also a slow acceptance by the state that some form of famine relief was necessary, although this may have been prompted more by fear of insurrection fueled by hunger than any humanitarian instinct.

If famines had become infrequent by the end of the eighteenth century malnutrition and undernutrition had not. In most West European nations, available supplies were insufficient to provide an adequate diet for all; the evidence on the height of young males suggests that a large proportion of the population were undernourished. In the nineteenth century available food supplies at the national level increased steadily; yet whilst the proportion of the population suffering from undernutrition probably declined, evidence of undernutrition and malnutrition persisted into the 1930s. Britain's available supplies probably first exceeded 3000 calories per capita per day in the 1850s or 1860s. Yet malnutrition still afflicted a significant proportion of the population in the 1930s. Undernutrition had declined for several reasons.

First was the progress of the knowledge of nutrition, amongst the population at large and amongst the medical profession. Important advances in the science of nutrition did not occur until this century; once the significance of proteins and vitamins became known it became possible to treat diseases such as rickets by providing vitamins in tablets. Second was the progress in providing clean drinking water and efficient sewage disposal combined with the discovery of germ theory and an understanding of the importance of hot water and soap. The reduction of the impact of contagious gastric diseases reduced child mortality. But it also reduced malnutrition and undernutrition, for vomiting children were undernourished children.

Third, and perhaps most important, has been the steady growth of real wages; admittedly historians have debated when precisely the benefits of the industrial revolution reached the workers, and undoubtedly there have been periods when real wages have fallen, none the less after 1850 the real income of most West Europeans rose. This enabled them to eat more of the staples – bread and potatoes – but also more of the more expensive foods such as vegetables, fruit, meat and milk. Slowly the extent of poverty was reduced. What is striking is how long it took for the real incomes of the poorest to reach a level where even they could buy a diet which ensured that they were free from malnutrition. In most of Western Europe this did not come until after the Second World War.

Whilst the growth of real incomes was a major cause of the decline of undernutrition in Western Europe – and poverty is now seen by many as the principal cause of hunger in the developing countries today – the paramount cause of hunger in pre-industrial Europe was the low productivity of agriculture; and the principal cause of the decline of hunger was the growth of productivity. In pre-industrial Europe crop yields were very low, as was output per capita; before 1800 average wheat yields exceeded 1000 kilograms per hectare in only the more intensively farmed areas, and between 15 and 25 per cent of that had to be retained to sow the following crop. Yields as low as this are found today only in some of the drier parts of the Middle East and North Africa.

With such low yields diets were bound to be meagre, stocks could not easily be built up and in years of drought and excessive rainfall or severe frosts yields were perilously low, as they were if there were outbreaks of plant disease. If yields per acre were low, so too were yields per capita; in most regions few people could be spared from food production for other trades, and from two-thirds to three-quarters of the population worked on the land. The great majority of the population were desperately poor, and spent between one-half and three-quarters of their income on food.

This changed as yields began to increase in the eighteenth and nine-teenth centuries. This increase allowed food output not only to keep up with total population growth but to increase available supplies per capita by 50 per cent or more in the nineteenth century. At the same time improvements in labour productivity allowed substantial migration from the countryside into the towns to provide the labour force for the new factory-based manufacturing industries. Indeed industrialization would have been difficult without this migration, for death rates in towns remained very high, often as high or higher than birth rates. Thus advances in agricultural productivity were a prerequisite for industrialization and the rising incomes that followed.

Clearly there was a substantial increase in European food supplies, mostly by increasing home production but also by imports. Although most writers have emphasized the importance of yield increases, there was a considerable expansion of the area in crops, both by reclaiming land hitherto uncultivated and by sowing crops upon the fallow. Indeed in England two-thirds of the extra output between 1750 and 1850 came from area expansion.[3] But by the end of the century arable land had reached its maximum and indeed in nearly every European country is now less than it was fifty years ago. Hence most of the increases in output in the last century have come from increased yields. In the eighteenth and nineteenth centuries higher yields were the result of the spread of mixed farming. Clover, improved grasses and root crops were sown on the fallow land of the open fields. The former provided grazing for livestock; far more important, clover fixed nitrogen in the soil and increased crop yields. Root crops, which included turnips, swedes, potatoes and sugar-beet, allowed weeding during growth, provided fodder and increased the number of livestock and so the supply of farmyard manure. Hence yields rose as the supply of nutrients was increased. This system spread slowly throughout Western Europe from the early eighteenth century; yet in the 1930s it had made little progress in éastern Europe.[4]

Nineteenth-century farmers could not improve their crops by selective breeding; they did, however, begin to grow hitherto unknown crops. The discovery of the Americas led to an exchange of plants; Europe acquired tobacco and the tomato, but especially maize and the potato. The latter increased greatly in importance in the nineteenth century; its calorific yield per hectare considerably exceeded that of wheat. Indeed some writers believe that it was the adoption of maize and potatoes that sustained Europe's great population growth.[5]

A combination of higher yields and area expansion increased Europe's food supplies in the nineteenth century. But available supplies were supplemented by imports. In the European-settled areas overseas a quite different path to agricultural progress was followed: with abundant land and low population densities the frontier of agricultural settlement swept across North America, Australasia and temperate South Africa. Few inputs were used; labour shortages led to the mechanization of production. Farmers had large farms and low rents and chose extensive rather than intensive production. Western Europe imported wheat, meat and dairy products from these countries, the volume increasing steadily after 1850. Europe also imported coarse grains and oilseeds, some of this from other parts of the world, to feed livestock. These food imports were well below the price of home-produced foods, and so allowed consumers in Europe to buy more food than they would have done if only home-produced food had been available, thus reducing the extent of malnutrition.

During the nineteenth and early twentieth centuries there were innovations in farming that heralded the great advances of the last fifty years. In the 1840s the first chemical fertilizers were produced in England, in the 1860s there were experiments with pesticides in France and the United States and in the 1890s selective herbicides were used in France. In 1900 Mendel's work in the genetics of plant breeding was rediscovered, and plant breeding institutes were rapidly established in many parts of Western Europe and North America. Just before the First World War Fritz Haber discovered a method of fixing the nitrogen in the atmosphere which allowed the production of cheap nitrogen fertilizers. A series of mechanical advances had been made in the United States – the reaper in the 1830s, a simple combine harvester in the 1850s and the tractor in the 1890s. These machines were subsequently much improved and by the 1920s and 1930s widely available if not widely adopted, for the poor economic conditions of the inter-war period delayed this. But in the post-war period the demand for food in the 1950s and the continued protection and subsidy of agriculture in Western Europe and North America has led to widespread adoption of these and subsequent agricultural innovations, with consequent dramatic increases in crop yields and output per capita. Indeed output has outrun consumer demand for food, and since the 1970s food surpluses have accumulated in the European Community and North America. However, public irritation at the high cost of protecting agriculture has lead to attempts to reduce output and so surpluses. By the early 1990s this was having some effect.

It is clear then that over the last 200 years the developed countries of the West have eliminated famine and reduced undernutrition and malnutrition to a very small proportion of the population. Nutritional diseases due to too much food have replaced nutritional diseases due to too little. It is equally clear that it is impossible to say that one factor alone has caused the reduction of hunger. Thus a growth in income has allowed the reduction of malnutrition; but the growth of income would not have occurred without industrialization, and industrialization required an increase in agricultural productivity and even more clearly, without an increase in output, there would have been no food for consumers to buy.

THE DEVELOPING COUNTRIES

In the period immediately after the Second World War most authorities took a gloomy view of the extent of hunger and the prospects for reducing it. This was not surprising: at the time Europe itself had acute food shortages; the first censuses taken after the war in Afro-Asia revealed alarming rates of increase, and the pre-war period had seen terrible famines in China and a decline in the output of food per capita in India. Nor did contemporaries foresee the power of modern agricultural technology, nor the possibility of economic growth in Asia or Africa. This pessimistic view has persisted; most prophets are still prophets of doom. Yet during the post-war period few of the gloomier predictions of the 1940s have been fulfilled. Population growth has been unprecedently rapid, and the rate of increase has only slackened in the last decade or so. Yet food output had kept ahead of population growth for the world as a whole, for the developing countries as a whole and in most individual countries in Latin America and Asia; it is only in tropical Africa that food output per capita has declined.

Estimates of the extent of hunger are easily criticized; none the less there seems to be some grounds for optimism. The incidence of famine is no longer as widespread as it was. India suffered a series of major famines in the nineteenth century, but after that of 1907–8 there was not another until the terrible Bengal famine of 1943. After independence there were problems in India when the monsoon failed in the 1960s, but no major crisis; indeed the last famine in the subcontinent occurred in Bangladesh in 1974. Famines had been chronic in China for centuries, and were particularly devastating in the nineteenth century, whilst in this century half a million died in 1920 and 2 million in 1929. The Communist government of 1948 was determined to reduce hunger both by increasing food production and introducing food rationing, so ensuring that all got an adequate diet. This policy seemed to be successful, and only in the late 1970s was it admitted that in 1958–61 there had been a major famine; it has since been estimated that 30 million people may have died then. The last thirty years have seen some undernutrition in China, but no major famines.[6] Indeed since 1960 – with the exception of the Bangladesh famine in 1974 – famines have been confined to tropical Africa, where before the 1960s there had been relatively few reports of major famines. Since the early 1970s there have been repeated famines in the Sahel, famine in Ethiopia in 1983–5, in many parts of southern Africa in the late 1980s and in Somalia in the early 1990s. These events have attracted much attention in the West, where television pictures have brought the horror to everyone's attention. Yet world-wide the frequency and intensity of famine has declined in the last fifty years.

The reasons for this decline are much the same as those which saw the decline in famines in Europe centuries ago. First, real transport costs, both overland and oversea, have fallen, allowing food to be dispatched to deficit areas more rapidly than in the past; it does not follow, of course, that food is always moved fast enough or in great enough quantities. Second, international organizations have organized famine relief and have encouraged Western governments to provide emergency food supplies and the more long-term food aid. Third, food output in every region except Africa has increased more rapidly than population. Consequently, whilst famines may be triggered by drought or locusts, their continuation is essentially man-made; civil war has been a major factor in causing famine in both Ethiopia and Somalia.

If famines have declined, or if their severity has become less because of prompter government intervention, chronic undernutrition and malnutrition persists in the developing countries. A recent FAO report estimates that in 1988–90 over 800 million, 20 per cent of the population of the developing countries, were chronically undernourished. However, both the numbers and the proportion were lower than in 1969–71.[7]

Over the longer period since 1950 it is difficult to be certain about the trend, for the many estimates made of the extent of hunger have very different methods of calculation, and for much of the period there are no reliable estimates for China, which contains one-fifth of the world's population. However, between 1950 and the later 1980s available supplies per capita rose by 20 per cent in the world and by 25 per cent in the developing countries; even in Africa, where food production per capita has been declining since the early 1960s, available supplies per capita have risen (table 3.5), albeit slightly. The number of people living in countries with less than 2200 calories per capita per day rose slightly between 1950 and 1987–9, but as a proportion of the world's population fell from 60 per cent to 27 per cent (table 3.6). A crude estimate (table 3.7) of those receiving less than 1.2 BMR indicates that between 1950 and 1986–8 the absolute numbers increased by 26 per cent, but the percentage of the population in the developing countries in this category fell from 34 per cent to 18 per cent; in Asia the actual numbers increased slightly – by 7 per cent – but fell as a proportion, in Latin America the numbers undernourished fell both absolutely and proportionally. It is in Africa that there has been a serious deterioration, the absolute numbers rising from 60 million in 1948–50 to 181 million in 1986–8. A final indirect measure is to compare the number of countries where food supplies exceeded national requirements: in the early 1960s few countries in the developing world had supplies above minimum requirements, by the mid 1980s such countries were mainly confined to South Asia and tropical Africa (figure 3.6).

It can be argued that these figures pay no regard to actual distribution of food amongst the population of a country, nor, more importantly, do they measure the clinical symptoms of undernutrition; they neglect the fact that the definition of undernutrition may be inaccurate. If it is borne in mind that the population of the developing countries increased by 140 per cent between 1950 and 1990, the comparatively small increase in the numbers undernourished suggests a considerable improvement.

The reasons for the relative decline in undernutrition in the developing countries are not fundamentally different from those that caused the decline in the now industrialized countries. There have been active attempts by governments in the developing countries to improve nutrition, both by encouraging agricultural development and by instituting nutrition programmes. Although infant and child mortality remains very high in developing countries when compared with Europe or North America, mortality rates have been declining (table 3.9): this suggests that the major killers of small children, diseases of the stomach and intestines, such as dysentery, are now of less significance due to the wider provision of sewage disposal and cleaner drinking water. Hence there are fewer children suffering from these disorders, and so fewer malnourished children. Indeed some believe that reducing the incidence of gastric disease amongst children may be as important a way of reducing the burden of malnutrition as increasing food supplies.

Modern writers have emphasized the importance of income in determining the extent of malnutrition. Numerous studies in the developing countries have shown a high positive correlation between income per capita and total calorie consumption and also with livestock products. There are clearly major differences in gross domestic product per capita between developed and developing countries. Hence, it is argued, the primary cause of malnutrition is poverty and the reduction of undernutrition depends upon the growth of incomes, in order that even the poorest groups have enough money to purchase a satisfactory diet. Although there are still marked differences in income between countries and although *it may* be that the gap between developed countries and developing countries has increased, the real gross domestic product per capita of all but a handful of developing countries has increased since 1950, in spite of a fall in some countries in the 1980s. Thus the ability of the poor to purchase food has improved and accounts for some of the relative decline in malnutrition.

In the 1950s and 1960s income distribution was rarely considered as a major cause of undernutrition. Most writers thought it a result of population growth outrunning food supplies; consequently a solution of the problem required a reduction in the rate of population growth and an increase in food output. But by the 1970s it was clear that the developing countries had been increasing their food output; indeed between the 1950s and 1980s output rose more rapidly in the developing than in the developed countries (table 5.5(a)). Because of the high rates of population growth in the former, output per capita rose more modestly (table 5.5(b)), but output per capita is now everywhere substantially above the 1950s, except in Africa (figures 5.2 and 5.3).

These conclusions led to a new emphasis in explanation; poverty was seen to be the major cause of undernutrition and the provision of available supplies of less significance. There is some justification for such a view; there are, for example, many developing countries with food supplies well above national requirements that have substantial numbers of undernourished people. But viewed historically the growth of food output has been paramount in the reduction of undernutrition. In 1950 most developing countries had very low supplies per capita per day; since then the population has risen from 1700 million to 3400 million, yet the proportion undernourished has fallen and the absolute numbers have risen but modestly. Clearly the expansion of food output has been crucial.

In 1950 many experts believed that most of the world's potential arable was already in cultivation, and so the area in crops could not be increased; and that crop yields in Europe were high, and unlikely to be further increased. Few thought a radical change in the traditional methods of Afro-Asia and Latin America to be likely. This has not proved so. Between 1950 and 1990 the arable area in the developing countries increased by 23 per cent; the sown area by much more, for everywhere – but particularly in Africa – that part of the arable in fallow has been reduced, and in Asia multiple cropping has increased. Nor have predictions about crop yields proved right; between 1950 and 1990 the yield of cereal crops in the developing countries has increased by 150 per cent.[8]

In Western Europe the expansion of the arable area made an important contribution to the increase of output in the eighteenth and nineteenth centuries; so it has in the developing countries, particularly in the less densely populated areas with large areas of potential arable. Since 1960 half the increase in cereal output in Africa came from area expansion, in Latin America one-third. Asia, where much of the cultivable land was already in crops in 1950, further increases in the sown area often depended upon irrigating drylands, which is costly. Hence most of the increase in output has come from high yields.

In the 1940s, few modern inputs – chemical fertilizers, modern varieties, pesticides or power driven machinery – were in use anywhere in the developing countries. In much of Africa and in many parts of Latin America plant nutrient content – and hence yields – were maintained by bush fallowing, and in neither region were animals and crops produced on the same farm, thus precluding the use of manure. In parts of East Asia high rice yields were obtained by the application of human and pig excreta and highly intensive cultivation. In most regions it proved possible to increase crop yields within these traditional systems, and indeed yields did increase, albeit very slowly in Africa, between the end of the Second World War and the mid 1960s. But however well farms were managed in the developing countries, it was not possible without the use of new inputs to match the production of land and labour found in the developed countries, nor would it be possible to achieve the high rates of productivity growth that rapid population growth required if food supplies per capita were to be maintained.

In the mid 1960s high yielding varieties of rice and wheat were introduced into parts of Asia; their successful growth required the use of chemical fertilizers, pesticides and irrigation. By the 1980s these had been successfully adopted in many parts of the developing world, but there were considerable regional variations in inputs, yields and the productivity of land and labour. Labour productivity was highest in Latin America and lowest in the Asian centrally planned economies (CPEs), reflecting differences in the use of tractors and other sources of power (table 12.1). On the other hand output per hectare was highest in the Far East and the Asian CPEs, reflecting differences in yields which in turn were influenced by the use of fertilizers and modern varieties. Africa falls last on virtually all counts; it is not surprising that this is the region where famine is still prevalent and where the highest proportion of the population are undernourished.

Productivity in the agriculture of the developing regions has increased dramatically in the last fifty years, but it still large well behind most parts of the developed world; there is thus a considerable 'slack' which can be taken up. Bringing levels of productivity up to present developed standards of farming would lead to a considerable increase in output. However, the reduction of undernutrition in the developing countries requires more than simply increasing future food supplies. Obviously it requires further improvements in income so that the poor can afford a better diet. But this hitherto has come only as a result of a shift from agricultural employment to manufacturing industry and services; in short future reduction of malnutrition depends upon future economic growth. Many of the other possible means of reducing undernutrition also seem to be dependent upon the general process of economic development. Thus a reduction in the prevalence of gastric disease amongst young children would reduce malnutrition, as would improvements in the status and education of women; improvements in transport and the marketing of food have a role to play, as have government programmes

Table 12.1 Productivity and inputs by major regions, 1988

	Latin America	Near East and North Africa	Far East	Africa	Asian CPEs
AGDP per capita of the agricultural labour force (dollars)	2076	1227	473	406	262
AGDP per hectare of agricultural land (dollars)	113	123	522	69	226
Tractors per 1000 agricultural workers	33	28	4	2	2
Percentage of power on farms from machinery[a]	19	12	2	2	n.d.
Kilogrammes of fertiliser per hectare of arable[b]	41	52	156	11	164
Percentage of all cereals sown with modern varieties[c]	27	19	49	10	56
Yields of all cereals, kilogrammes per hectare	2079	1604	2034	1052	3860
Percentage of all arable irrigated	9	24	28	4	44

CPEs, centrally planned economies.
AGDP, Agricultural Gross Domestic Product.
[a] 1980.
[b] 1984–6.
[c] 1982–3.
Sources: World Bank, *World Development Report 1990*, Oxford, 1990; FAO, *Production Yearbook 1989*, vol. 43, Rome, 1990; N. Alexandratos, J. Bruinsma and J. Hrabovsky, 'Power inputs from labour, draught animals and machines in the agriculture of the developing countries', *European Review of Agricultural Economics*, vol. 9. 1982, pp. 127–55

of nutritional supplementation. Hence although this book has concerned itself primarily with the growth of food output, it has to be recognized that the reduction of undernutrition is a function of that series of interrelated social and economic changes sometimes described as modernization.

Notes

1 Introduction

1 Sir William Crookes, *The Wheat Problem*, London, 1917; Yves Guyot, 'The bread and meat of the world', *American Statistical Association*, 67–8, 1904, pp. 79–119; G. P. Roorbach, 'The world's food supply', *Annals of the American Academy of Political and Social Science*, 74, 1917, pp. 1–13; League of Nations, *The Problem of Nutrition*, 3 vols, Geneva, 1936.
2 G. C. L. Bertram, 'Population trends and the world's resources', *Geographical Journal*, 107, 1946, pp. 191–210.
3 K. McQuillan, 'Common themes in Catholic and Marxist thought on population and development', *Population and Development Review*, 5, 1979, pp. 689–99.
4 W. Vogt, *Road to Survival*, London, 1949; W. Paddock and P. Paddock, *Famine 1975*, London, 1967; D. H. Meadows and D. I. Meadows, *The Limits to Growth: A Report for the Club of Rome on the Predicament of Mankind*, London, 1972.

2 The extent of hunger

1 Sir John Boyd Orr, 'The food problem', *Scientific American*, 183, 1950, pp. 11–15; J. de Castro, *The Geography of Hunger*, London, 1952, p. 16.
2 Ministry of Agriculture, *Manual of Nutrition*, London, 1970, p. 13. The term 'calorie' is used in this book, as is conventional, although strictly speaking the units are kilocalories.
3 (FAO/WHO) *Energy and Protein Requirements; Report of a Joint FAO/WHO Ad Hoc Expert Committee*, Geneva, 1973, p. 107.
4 T. N. Srinivasan, 'Measuring malnutrition', *Ceres*, 16 (2), 1983, pp. 23–7; P. V. Sukhatme and S. Margen, 'Autoregulatory homeostatic nature of energy balance', *American Journal of Clinical Nutrition*, 35, 1982, pp. 355–65; W. Edmundson, 'Individual variations in basal metabolic rate and mechanical work efficiency in East Java', *Ecology of Food and Nutrition*, 8, 1979, pp. 189–95.

5   FAO/WHO, *Energy and Protein Requirements*, p. 29.
6   Sukhatme and Margen, 'Autoregulatory homeostaric nature'; Edmundson, 'Individual variations', Srinivasan, 'Measuring malnutrition'.
7   Ministry of Agriculture, *Manual of Nutrition*, p. 18.
8   D. M. Hegsted, 'Protein calorie malnutrition', *American Scientist*, **66**, 1978, pp. 61–5.
9   D. S. McLaren, 'The great protein fiasco', *The Lancet*, **ii**, 1974, pp. 93–6.
10 P. R. Payne, 'Proteins in human nutrition; nutritional requirement and social needs', *Folia Veterinaria Latina*, **6**, 1976, pp. 23–33.
11 P. V. Sukhatme, 'Human calorie and protein needs and how far they are satisfied today', in B. Benjamin, P. R. Cox and J. Peel (eds), *Resources and Population*, London, 1973, pp. 25–43; F. Aylward and M. Jul, *Protein and Nutrition Policy in Low Income Countries*, London, 1975, p. 32; FAO, 'The food situation and the child: an overview', *Food and Nutrition*, 5, 1979, pp. 4–8.
12 McLaren, 'The great protein fiasco'; J. C. Waterlow and P. R. Payne, 'The protein gap', *Nature*, **258**, 1975, pp. 113–17.
13 R. Passmore, B. Nicol and M. Rao, *Handbook on Human Nutritional Requirements*, Geneva, 1974.
14 J. M. Bengoa, 'Prevention of protein–calorie malnutrition', in R. E. Olsen (ed.), *Protein–Calorie Malnutrition*, London, 1975; J. Mayer, 'The dimensions of human hunger', *Scientific American*, **235**, 1976, pp. 40–9; J. Yudkin, 'Some basic principles of nutrition', in D. S. Miller and D. J. Oddy (eds), *The Making of the Modern British Diet*, London, 1976, pp. 196–203.
15 M. C. Latham, 'Nutrition and infection in national development', *Science*, **188**, 1975, pp. 561–5.
16 R. Martorell, 'Interrelationships between diet, infectious disease and nutritional status', in L. S. Greene and F. E. Johnston (eds), *Social and Biological Predictors of Nutritional Status, Physical Growth, and Neurological Development*, London, 1980, pp. 81–106; N. S. Scrimshaw, 'Interactions of malnutrition and infection: advances in understanding', in Olsen, Protein–Calorie Malrutrition, pp. 353–67; N. S. Scrimshaw, 'World nutritional problems', in L. Newman (ed.), *Hunger in History, Food Shortage, Poverty and Deprivation*, Oxford, 1990, pp. 353–73.
17 Hegsted, 'Protein calorie malnutrition', pp. 61–5; R. W. Wenlock, 'Endemic malaria, malnutrition and child deaths', *Food Policy*, **6** (2), 1981, pp. 105–12; L. J. Mata, R. A. Kranial, J. T. Urrutia and B. Garcia, 'Effect of infection on food intake and the nutritional state: perspectives as viewed from the village', *American Journal of Clinical Nutrition*, **30**, 1977, pp. 1215–27; D. Jameson and A. Piazza, 'China's food and nutritional planning', in J. P. Gittinger, J. Leslie and C. Hoisington (eds), *Food Policy, Integrating Supply, Distribution and Consumption*, Baltimore, Md., 1987, pp. 467–84.
18 C. M. Fillmore and M. A. Hussain, 'Agriculture and anthropometry: assessing the nutritional impact', *Food and Nutrition*, **10**, 1984, pp. 2–14.
19 G. H. Beaton and J. M. Bengoa, 'Nutrition and health in perspective: an introduction', in G. H. Beaton and J. M. Bengoa (eds), *Nutrition in*

Preventive Medicine: The Major Deficiency Syndromes, Epidemiology and Approaches to Control, Geneva, 1971, p. 45; C. Gopalan, 'Protein versus calories in the treatment of protein calorie malnutrition: metabolic and population studies in India', in Olsen, *Protein–Calorie Malnutrition*, pp. 329–41.

20 J. M. Bengoa, 'The state of world nutrition', in M. Rechcigl, Jr., *Man, Food and Nutrition*, Cleveland, Ohio, 1973, pp. 1–13; Mayer, 'The dimensions of human hunger'; Bengoa, 'Prevention of protein–calorie malnutrition'; Passmore *et al.*, *Human Nutritional Requirements*; J. D. Haas and G. G. Harrison, 'Nutritional anthropology and biological adaptation', *Annual Review of Anthropology*, 6, 1977, pp. 69–101.

21 FAO, *The Fourth World Food Survey*, Rome, 1977, p. 110; United Nations Committee on Co-ordination-Sub-committee on Nutrition, *First Report on the World Nutrition Situation*, New York, 1987; M. C. Latham, 'Strategies for the control of malnutrition and the influence of the nutritional sciences', *Food and Nutrition*, 10, 1984, pp. 5–31.

22 M. Murioz de Chavez, 'Malnutrition: socioeconomic effects and policies in developing countries', in P. B. Pearson and J. R. Greenwell (eds), *Nutrition, Food and Man*, Tucson, Ariz., 1980, pp. 38–45.

23 E. M. Demaeyer, 'Clinical manifestations of malnutrition', in D. N. Walcher, N. Kretchmer and H. L. Barnett (eds), *Food, Man and Society*, New York, 1976.

24 K. M. Cahill, 'The clinical face of famine in Somalia', in K. M. Cahill (ed.), *Famine*, New York, 1982.

25 Cahill, 'The clinical face' Demaeyer, 'Chinical manifestations'.

26 J. M. Bengoa, 'Recent trends in the public health aspects of protein calorie malnutrition', *WHO Chronicle*, 24, 1970, pp. 552–61.

27 J. Haaga, C. Kennick, K. Test and J. Mason, 'An estimate of the prevalence of child malnutrition in developing countries', *World Health Statistics Quarterly*, 38, 1985, pp. 331–47.

28 Bengoa, 'Prevention of protein–calorie malnutrition'.

29 D. S. Miller, 'Nutrition surveys', in Miller and Oddy, *Modern British Diet*, pp. 202–13.

30 FAO/WHO, *Energy and Protein Requirements*; FAO/WHO *Energy and Protein Requirements*, WHO Technical Report No. 724, Geneva, 1985; see also W. P. T. James and E. C. Schofield, *Human Energy Requirements: a manual for planners and nutritionists*, Oxford, 1990.

31 Edmundson, 'Individual variations'; Sukhatme and Margen, 'Autoregulatory homeostatic nature'; N. S. Scrimshaw and L. Taylor, 'Food', *Scientific American*, 243, 1980, pp. 74–84.

32 B. M. Nicol, 'Causes of famine in the past and in the future', in G. Blix, Y. Hofvander and B. Vahlquiust *Famine: A Symposium*, Uppsala, Swedish Nutrition Foundation, 1971; John Bennet, *The Hunger Machine: The Politics of Food*, London, 1987, p. 9.

33 FAO, *Production Yearbooks*, since 1961; FAO, *World Food Survey*, Washington, D.C., 1946; FAO, *The Second World Food Survey*, Rome, 1952; FAO, *The Fourth Survey*; K. Becker, 'Food balance sheets', *FAO Quarterly Bulletin of Statistics*, 4, 1988, pp. iii–v.

34 M. K. Bennett, 'Longer and shorter views of the Malthusian prospect', *Food Research Institute Studies*, 4, 1963, pp. 3–12; W. H. Calloway, 'World calorie/protein needs', in Pearson and Greenwell, *Nutrition, Food and Man*, pp. 82–7; C. Clark and J. Boyd Turner, 'World population growth and future food trends', in Rechcigl, *Man, Food and Nutrition*, pp. 55–77; R. W. Hay, 'The statistics of hunger', *Food Policy*, 3, 1978, pp. 243–55; T. T. Poleman, *Quantifying the Nutrition Situation in Developing Countries*, Cornell Agricultural Economics Staff Paper No. 79–33, 1979; T. T. Poleman, 'World food: myth and reality', in R. Sinha (ed.), *The World Food Problem: Consensus and Conflict*, Oxford, 1978, pp. 383–94.

35 C. Geissler and D. Miller, 'Nutrition and GNP: a comparison of problems in Thailand and the Philippines', *Food Policy*, 7, 1982, pp. 191–206; J. Katzmann, 'Besoins alimentaires et potentialités des pays en voie de développement', *Mondes Développées*, 29–30, 1980, pp. 53–6; V. Smil, *Energy, Food, Environment*, Oxford, 1987, p. 150.

36 FAO/WHO, *Energy and Protein Requirements*, p. 78.

37 FAO, *The Fourth Survey*, 1977, pp. 77–80.

38 L. M. Li, 'Feeding China's one billion: perspectives from history', in Cahill, *Famine*; V. Smil, 'Food production and quality of diet in China', *Population and Development Review*, 12, 1986, pp. 25–45.

39 J. C. Waterlow, 'Childhood malnutrition – the global problem', *Proceedings of the Nutrition Society*, 38, 1979, pp. 1–9.

40 S. Reutlinger and H. Alderman, 'The prevalence of calorie deficit diets in developing countries', *World Development*, 8, 1980, pp. 399–411.

41 W. C. Edmundson and P. V. Sukhatme 'Food and work; poverty and hunger', *Economic Development and Cultural Change*, 38, 1990, pp. 263–80; J. N. Srinivasan, 'Undernutrition: extent and distribution of its incidence', in A. Maunder (ed.), *Proceedings of the Nineteenth International Conference of Agricultural Economists, Malaga, Spain, 1985*, Aldershot, 1986, pp. 199–214; N. Edirisinghe and T. T. Poleman, 'Behavioural thresholds as indicators of perceived dietary adequacy or inadequacy', *Cornell International Agricultural Economics Study*, Ithaca, N. Y., 1983; S. R. Osman, 'Controversies on nutrition and their implications for the economics of food', *Wider Working Papers, World Institute of Development Economics Research*, Helsinki, 1987.

42 FAO, *The State of Food and Agriculture 1987–88*, Rome, 1988, p. 58.

3 A short history of hunger

1 Carlo M. Cipolla, *Before the industrial Revolution: European Society and Economy 1000–1700*, London, 1976, p. 31.

2 Wilhelm Abel, *Agricultural Fluctuations in Europe from the Thirteenth to the Twentieth Centuries*, London, 1980, p. 32; F. Braudel, *Capitalism and Material Life 1400–1850*, London, 1974, p. 87; D. Perkins, *Agricultural Development in China, 1368–1968*, Edinburgh, 1969; Food and Agriculture Organization (FAO), *Production Yearbook 1988*, vol. 42, Rome, 1989.

3   Abel, *'Agricultural Fluctuations*, pp. 71, 255; Braudel, *Capitalism*, pp. 67, 128–30; B. Bennassar and J. Goy, 'Contribution a l'histoire de la consommation alimentaire du XIVe siècle', *Annales ESC*, **30**, 1975, pp. 402–30.

4   H. S. Lucas, "The great European famine of 1315, 1316 and 1317", *Speculum*, **5**, 1930, pp. 343–77; M. Bergman, 'The potato blight in the Netherlands and its social consequences 1845–1847', *International Review of Social History*, **12**, 1967, pp. 390–431; A. B. Appleby, 'Epidemics and famine in the Little Ice Age', *Journal of Interdisciplinary History*, **10**, 1979, pp. 643–63; J. D. Post, *The Last Great Subsistence Crisis in the Western World*, London, 1977.

5   M. W. Flinn, 'The stabilisation of mortality in pre-industrial Western Europe', *Journal of European Economics History*, **32**, 1974, pp. 1289–328.

6   J. Meuvret, 'Demographic crisis in France from the sixteenth to the eighteenth century', in D. V. Glass and D. E. C. Eversley (eds), *Population in History: Essays in Historical Demography*, London, 1965, pp. 507–22; J. Stevenson, 'Food riots in England, 1792–1818', in R. Quinault and J. Stevenson (eds), *Popular Protest and Public Order: Six Studies in British History 1790–1920*, London, 1974, pp. 33–74; C. Walford, 'On the famines of the world, past and present', *Journal of the Statistical Society*, **41**, 1878, pp. 433–526.

7   J. Hemardinquer, *Pour une Histoire de l'alimentation*, Paris, 1970.

8   J. C. Toutain, *La Consommation Alimentaire en France de 1789 à 1964*, *Economies et Sociétés*, vol. 5, Paris, 1971.

9   C. Lis and H. Soly, 'Food consumption in Antwerp between 1807 and 1859; a contribution to the standard of living debate', *Economic History Review*, **30**, 1977, pp. 460–81.

10 Lis and Soly, 'Food consumption'; Toutain, *La Consommation*, pp. 79–83; M. Drake, 'Norway', in W. R. Lee (ed.), *European Demograph and Economic Growth*, London, 1979, p. 293; D. J. Oddy, 'The health of the people', in T. Barker and M. Drake (eds), *Population and Society in Britain 1850–1980*, London, 1982, pp. 121–32; W. A. McKenzie, 'Changes in the standard of living in the United Kingdom, 1860–1914', *Economica*, **1**, 1921, pp. 211–30.

11 M. Aymard, 'Towards the history of nutrition: some methodological remarks', in R. Forster and O. Ranum (eds), *Food and Drink in History: Selections from the Annales*, Baltimore, Md., 1979, pp. 1–16; H. J. Teuteberg, 'The general relationships between diet and industrialization', in E. Forster and R. Forster (eds), *European Diet from Preindustrial to Modern Times*, London, 1975, pp. 61–110; Toutain, *La Consommation*.

12 P. Bairoch, 'Ecarts internationaux des niveaux de vie avant la revolution industrielle', *Annales ESC*, **34**, 1979, pp. 145–71.

13 C. Dauphin and P. Pezerat, 'Les consommations populaires dans la seconde moitié du XIXe siècle à travers les monographies de l'école de Le Play', *Annales ESC*, **30**, 1975, pp. 537–52.

14 T. C. Barker, D. J. Oddy and J. Yudkin, *The Dietary Surveys of Dr. Edward Smith, 1862–3: A New Assessment*, Department of Nutrition, Queen

Elizabeth College, University of London, Occasional Paper no. 1, London, 1970; J. C. McKenzie, 'The composition and nutritional value of diets in Manchester and Dukinfield, 1841', *Lancashire and Cheshire Antiquarian Society Transactions*, 72, 1962, pp. 126–39; B. E. Supple, 'Income and demand, 1860–1913', in R. Floud and D. McCloskey (eds), *The Economic History of Britain since 1700. Volume 2: 1860 to the 1970s*, Cambridge, 1981, p. 132; D. J. Oddy, 'Food in nineteenth century England: nutrition in the first urban society', *Proceedings of the Nutrition Society*, 29, 1970, pp. 150–7.

15 John Burnett, *Plenty and Want: A Social History of Diet in England from 1815 to the Present Day*, Harmondsworth, 1968, pp. 271, 301, 303, 307; D. J. Oddy, 'The health of the people'; F. Le Gros Clark and R. M. Titmus, *Our Food Problem and its Relation to our National Defences*, Penguin, 1939.

16 Sir John Boyd Orr, *Food, Health and Income*, London, 1937; Le Gros Clark and Titmus, *Our Food Problem*, p. 125; League of Nations, *The Problem of Nutrition*, vol. 1: *Interim Report of the Mixed Committee on the Problem of Nutrition;* vol. 3: *Nutrition in Various Countries*, Geneva, 1936; J. de Castro, *The Geography of Hunger*, London, 1952.

17 J. M. Tanner, 'Earlier maturation in man', *Scientific American*, 218, 1968, pp. 21–7; G. H. Bruntland, K. Liestøllesdal and L. Walløe, 'Height, weight and menarcheal age of Oslo school children during the last 60 years', *Annals of Human Biology*, 7, 1980, pp. 307–22; B. Ljung, A. Bergsteen-Brucefors and G. Lindgren, 'The secular trend in physical growth in Sweden', *Annals of Human Biology*, 1, 1974, pp. 245–56; R. Floud and K. W. Wachter, 'Poverty and physical stature: evidence on the standard of living of London boys, 1770–1870', *Social Science History*, 6, 1982, pp. 422–52.

18 Tanner, 'Earlier maturation'; K. Liestol, 'Social conditions and menarcheal age: the importance of early years of life', *Annals of Human Biology*, 9, 1982, pp. 521–37.

19 Tanner, 'Earlier maturation'; R. W. Fogel, S. L. Engerman and J. Trussell, 'Exploring the uses of data in height: the analysis of long term trends in nutrition, labour welfare and labour productivity', *Social Science History*, 6, 1982, pp. 401–21; R. W. Fogel, *Nutrition and the Decline in Mortality since 1700: some Additional Preliminary Findings*, National Bureau of Economic Research Work Paper No. 1802, Cambridge, 1986; L. G. Sandberg and R. H. Steckel 'Overpopulation and malnutrition rediscovered; hard times in nineteenth century Sweden', *Explorations in Economic History*, 25, 1988, pp. 1–19; J. Komlos, 'The height and weight of West Point cadets; dietary change in antebellum America', *Journal of Economic History*, 47, 1987, pp. 897–927.

20 W. R. Aykroyd, 'Nutrition and mortality in infancy and early childhood: past and present relationships', *American Journal of Clinical Nutrition*, 24, 1971, pp. 480–7; S. H. Preston and E. van de Walle, 'Urban French mortality in the nineteenth century', *Population Studies*, 32, 1978, pp. 275–97.

21 G. Fridlizius, 'Sweden', in W. R. Lee (ed.), *European Demography and*

Economic Growth, London, 1979, pp. 284–318; O. Andersen, 'Denmark', in Lee, *European Demography*, p. 112; R. I. Woods, 'The structure of mortality in mid-nineteenth-century England and Wales', *Journal of Historical Geography*, **8**, 1982, pp. 373–94; P. Deprez, 'The Low Countries', in Lee, *European Demography*, p. 281; Preston and van de Walle, 'Urban French mortality'.

22 T. McKeown, 'Food, infection and population', *Journal of Interdisciplinary History*, **14**, 1983, pp. 227–47; A. G. Carmichael, 'Infection, hidden hunger and history', *Journal of Interdisciplinary History*, **14**, 1983, pp. 249–64; M. Livi-Bacci, 'The nutrition-mortality link in past times: a comment', *Journal of Interdisciplinary History*, **14**, 1983, pp. 293–6.

23 M. K. Bennett, 'International contrasts in food consumption', *Geographical Review*, **31**, 1941, 365–76.

24 Sir John Boyd Orr, 'The food problem', *Scientific American*, **183**, 1950, pp. 11–15.

25 S. Mazumdar, 'Realistic food goals for Africa', *Ceres*, **13**, 1980, pp. 36–41.

26 United Nations Administrative Committee on Co-ordination–Subcommittee on Nutrition, *First Report on the World Nutrition Situation*, New York, 1987; World Health Organization, *Update on the Nutrition Situation. Recent trends on nutrition in 33 Countries*, Geneva, 1989.

27 G. H. Beaton and J. M. Bengoa, 'Practical population indicators of health and nutrition', in G. H. Beaton and J. M. Bengoa (eds), *Nutrition in Preventive Medicine*, Geneva, 1971.

28 Le Gros Clark and Titmus, *Our Food Problem*; F. B. Smith, *The People's Health 1830–1910*, London, 1979, pp. 17, 178–83; D. J. Oddy, 'Working class diets in late nineteenth century Britain', *Economic History Review*, **23**, 1970, pp. 314–23.

4 Population and poverty

1 FAO, *World Food Survey*, Washington, D.C., 1946; FAO, *The Second World Food Survey*, Rome, 1952.

2 D. Grigg, 'Modern population growth in historical perspective', *Geography*, **67**, 1982, pp. 97–108.

3 United Nations, *Demographic Yearbook 1980*, New York, 1982.

4 S. H. Preston, 'The changing relationship between mortality and level of economic development', *Population Studies*, **29**, 1978, pp. 231–48; S. H. Preston, 'Causes and consequences of mortality declines in less developed countries during the twentieth century', in R. A. Easterlin (ed.), *Population and Economic Change in Developing Countries*, Chicago, Ill., 1980, pp. 289–360; W. Parker Maudlin, 'Population trends and prospects', *Science*, **209**, 1980, pp. 148–57; D. R. Gwatkins, 'Indications of change in developing country mortality trends: the end of an era', *Population and Development Review*, **6**, 1980, pp. 615–44.

5 S. Kuznets, *Population, Capital and Growth; Selected Essays*, 1973, pp.

95-102; D. Kirk, 'World population and birth rates: agreements and disagreements', *Population and Development Review*, 5, 1979, pp. 387-404; R. H. Cassen, 'Current trends in population change and their causes', *Population and Development Review*, 4, 1978, pp. 331-53; Y. Blayo and J. Verron, 'La fécondité dans quelques pays d'Asie oriental', *Population*, 32, 1977, pp. 945-7; D. F. S. Fernando, 'Recent fertility decline in Ceylon', *Population Studies*, 26, 1973, pp. 445-54; D. F. S. Fernando, 'Changing nuptiality patterns in Sri Lanka 1901-1971', *Population Studies*, 29, 1975, pp. 179-90.

6   A. J. Coale, 'Population trends, population policy and population studies in China', *Population and Development Review*, 7, 1981, pp. 85-7; United Nations, *Demographic Yearbook 1980*.

7   J. Bourgeois Pichat, 'Recent demographic changes in Western Europe', *Population and Development Review*, 7, 1981, pp. 19-42.

8   P. Bairoch, *The Economic Development of the Third World since 1900*, London, 1975, p. 6.

9   D. Grigg, *Population Growth and Agrarian Change; An Historical Perspective*, Cambridge, 1980, p. 238; J. Ho, 'La evolution de la mortalité en Amerique Latine', *Population*, 25, 1970, pp. 103-6.

10   J. D. Durand, 'Historical estimates of world population; an evaluation', *Population and Development Review*, 3, 1977, pp. 253-96.

11   D. Grigg, *Population Growth*, 1980, p. 238.

12   J. R. Harrison, 'Third World Incomes before World War I: some comparisons', *Explorations in Economic History*, 25, 1988, pp. 323-36.

13   World Bank, *World Development Report 1990*, Washington, D.C., 1990, p. 26; United Nations, *Human Development Report 1990*, Oxford, 1990, p. 18.

14   K. Griffin, *The Political Economy of Agrarian Change*, Oxford, 1974; A. de Janvry, *The Agrarian Question and Reformism in Latin America*, Baltimore, Md., 1981; International Labour Office, *Poverty and Landlessness in Rural Asia*, Geneva, 1977; World Bank, *World Development Report 1980*, Washington, D.C., 1980, pp. 33-62; M. S. Ahluwalia, N. G. Carter and H. B. Chenery, 'Growth and poverty in developing countries', *Journal of Development Economics*, 6, 1979, pp. 299-345; P. Streeten, *First Things First: Meeting Basic Human Needs in Developing Countries*, Oxford, 1980; I. Adelman and C. T. Morris, *Economic Growth and Social Equality in Developing Countries*, Stanford, 1973; H. Chenery, M. S. Ahluwalia, C. L. G. Bell, J. H. Duby and R. Jolly, *Redistribution with Growth*, Oxford, 1974; World Bank, *World Development Report 1981*, Washington, D.C., 1981, pp. 134-5; R. Gaitia, 'Impoverishment, technology and growth in rural India', *Cambridge Journal of Economics*, 11, 1987, pp. 23-46; V. Kirkpatrick and G. T. Harris, 'A note on trends in poverty and inequality in rural Asia, 1950-1985', *Journal of Contemporary Asia*, 19, 1989, pp. 324-30.

15   N. Alexandratos (ed.), *World Agriculture towards 2000. An FAO Study*, London, 1988, p. 220; World Bank, *Report 1981*, pp. 30-40.

16   World Bank, *Report 1990*, pp. 40-1, 45; V. G. Bhatia, 'Asia and Pacific developing economies; performance and issues', *Asian Development Review*, 6, 1988, pp. 1-21.

5 The growth of world food output

1 G. B. Cressey, 'Land for 2.4 billion neighbours', *Economic Geography*, **29**, 1953, pp. 1–9; R. M. Salter, 'World soil and fertilizer needs in relation to food needs', *Science*, **105**, 1947, pp. 533–8.

2 Sir W. Crookes, *The Wheat Problem*, London, 1899; G. C. Anderson, 'An agricultural view of the world population – food crisis', *Journal of Soil and Water Conservation*, **27**, 1972, pp. 52–6; L. R. Brown and E. P. Eckholm, *By Bread Alone*, London, 1975; Paul and Anne Ehrlich, *Population Explosion*, London, 1990.

3 E. Le Roy Ladurie and J. Goy, *Tithe and Agrarian History from the Fourteenth to the Nineteenth Century*, Cambridge, 1982.

4 B. H. Slicher van Bath, 'Agriculture in the vital revolution', in C. H. Wilson and E. E. Rich (eds), *Cambridge Economic History of Europe*, vol 4: *The Economic Organization of Early Modern Europe*, Cambridge, 1977; P. Deane and W. A. Cole, *British Economic Growth, 1688–1959*, Cambridge, 1962, pp. 65, 78; J. C. Toutain, *Le Produit de l'Agriculture Francais*, vol. 1, Paris, 1961, pp. 213–15; T. R. Malthus, *An Essay on the Principle of Population*, London, 1798.

5 Toutain, *Le Produit*, vol. 2, pp. 127–9; Deane and Cole, *British Economic Growth*, p. 170; T. W. Fletcher, 'Drescher's index: a comment', *Manchester School of Economic and Social Studies*, **23**, 1955, p. 181; W. R. Lee, 'Primary sector output and mortality changes in early XIXth century Bavaria', *Journal of European Economic History*, **6**, 1977, pp. 155–62; M. R. Haines, 'Agriculture and development in Prussian Upper Silesia, 1846–1913', *Journal of Economic History*, **42**, 1982, pp. 355–84; M. Towne and W. D. Rasmussen, 'Farm gross product and gross investment in the nineteenth century', in National Bureau of Economic Research, *Trends in the American Economy in the Nineteenth Century. Studies in Income and Wealth*, vol. 24, Princeton, N.J., 1960, p. 260.

6 Paul Bairoch, *The Economic Development of the Third World since 1900*, London, 1975; G. Blyn, *Agricultural Trends in India, 1891–1957: Output, Availability, and Productivity*, Philadelphia, Pa, 1966; D. H. Perkins, *Agricultural Development in China, 1368–1968*, Edinburgh, 1969.

7 L. R. Brown, *The Changing World Food Prospect: the Nineties and Beyond*, World Watch Paper 85, Washington, D.C., 1985, p. 46.

8 FAO, *Production Yearbooks*, Rome, various years; *The Agricultural Production Index*, Rome, 1986.

9 FAO, *The Index*.

10 J. L. Simon, 'Resources, population, environment: an oversupply of false bad news', *Science*, **208**, 1980, pp. 1431–7.

11 Slicher van Bath, 'Agriculture', p. 81; M. Overton, 'Estimating crop yields from probate inventories; an example from East Anglia, 1585–1735', *Journal of Economic History*, **39**, 1975, pp. 363–78; D. B. Grigg, *The Dynamics of Agricultural Change: The Historical Experience*, London, 1982, pp. 130, 172–5, 184–9.

12   G. P. H. Chorley, 'The agricultural revolution in northern Europe, 1750–
     1880: nitrogen, legumes and crop productivity', *Economic History Review*,
     34, 1981, pp. 71–93.
13   G. E. Mingay, *The Agricultural Revolution: Changes in Agriculture 1650–1880*,
     London, 1977; D. Grigg, *Dynamics*, p. 187.
14   FAO, *Roots, Tubers and Plantains in Food Security*, FAO Economic and Social
     Development Paper No. 79, Rome, 1989.
15   FAO, *Production Yearbook 1989*, vol. 43, Rome, 1990.
16   World Bank, *Accelerated Development in Sub-Saharan Africa: An Agenda
     for Action*, Washington, D.C., 1981; C. Christensen et al., *Food Problems
     and Prospects in Sub-Saharan Africa: The Decade of the 1980s*, International
     Economics Division, Economic Research Service, United States Depart-
     ment of Agriculture, Washington, D.C., 1981; FAO, *The State of Food and
     Agriculture 1978*, Rome, 1979, pp. 2–32.

6   The expansion of the world's arable land

1   D. Grigg, 'The growth and distribution of the world's arable land, 1870–
    1970', *Geography*, **59**, 1974, pp. 104–10; see also J. F. Richards, 'Global
    patterns of land conservation', *Environment*, **26**, 1984, pp. 6–38, and P.
    Buringh and R. Dudal, 'Agricultural land use in space and time', in M. G.
    Wolman and F. G. A. Fournier (eds), *Land Transformation in Agriculture*,
    Chichester, 1987, pp. 9–44.
2   D. H. Jansen, 'Tropical agroecosystems', *Science*, **183**, 1974, 1212–19; Jen-
    Hu Chang, 'The agricultural potential of the humid tropics', *Geographical
    Review*, **58**, 1968, 33–61.
3   FAO, *Provisional Indicative World Plan for Agricultural Development*, vol. 1,
    Rome, 1970, p. 45.
4   FAO, *Provisional World Plan*, p. 44.
5   D. J. Fox, 'Mexico', in H. Blakemore and C. T. Smith (eds), *Latin
    America: Geographical Perspectives*, London, 1983, p. 41; FAO, *Provisional
    World Plan*, pp. 44–5.
6   P. Buringh, 'Food production potential of the world', in R. Sinha (ed.),
    The World Food Problem, London, 1978, pp. 477–85.
7   D. G. Dalrymple, *Survey of Multiple Cropping in Less Developed Nations*,
    Foreign Economic Development Service, US Department of Agriculture,
    Washington DC, 1971; D. J. Andrews and A. H. Kassav, 'The importance
    of multiple cropping in increasing world food supplies', in R. I. Papendick,
    P. A. Sanchez, G. B. Triplett (eds), *Multiple Cropping*, American Society
    of Agronomy Special Publication No. 27, Madison, Wis., 1976, pp. 1–10;
    Fox, 'Mexico'.
8   H. G. Nasr, 'Multiple cropping in some countries of the Middle East',
    in Papendick et al., *Multiple Cropping*, pp. 117–27; Dalrymple, *Survey of
    Multiple Cropping*.
9   E. Dayal, 'Impact of irrigation expansion on multiple cropping in India',
    Tijdschrift voor Economische en Sociale Geografie, **68**, 1971, pp. 100–9;

Dalrymple, *Survey of Multiple Cropping*; S. Ishikama, 'China's economic growth since 1949 – an assessment', *China Quarterly*, 94, 1983, pp. 242–81.

10 N. D. Gulhati, *Irrigation in the World: A Global Review*, New Delhi, 1955, p. vii.

11 L. R. Brown and E. Eckholm, *By Bread Alone*, London, 1975, p. 94; Gulhati, *Irrigation*; FAO, *Production Yearbook 1989*, vol. 43, Rome, 1990; S. H. Wittwer, 'Food production: technology and the resource base', *Science*, 188, 1975, pp. 579–84.

12 W. C. Beets, *Multiple Cropping and Tropical Farming Systems*, Boulder, Colo., 1982, pp. 7, 8, 10, 43, 63; M. A. Altieri; *Agroecology: the Scientific Basis of Alternative Agriculture*, Boulder, Colo., 1987, p. 116.

13 T. J. Goering, *Agricultural Land Settlement*, World Bank Issues Paper, Washington, D.C., 1978; L. R. Brown, 'Food growth slowdown: danger signal for the future', in J. P. Gittinger, J. Leslie and C. Hoisington (eds), *Food Policy: Integrating Supply, Distribution, Consumption*, Baltimore, Md., 1987, pp. 89–105.

14 Tunku Shamsul Bahrein, 'Development planning: land settlement policies and practices in South East Asia', in R. J. Pryor (ed.), *Migration and Development in South East Asia; a Demographic Perspective*, Kuala Lumpur, 1979, pp. 295–304; R. Ng, 'Land settlement projects in Thailand', Geography, 53, 1968, pp. 179–82; B. H. Farmer, *Agricultural Colonization in India since Independence*, London, 1974; R. C. Eidt, 'Pioneer settlement in Eastern Peru', *Annals of the Association of American Geographers*, 52, 1962, pp. 255–78.

15 R. D. Hill, *Agriculture in the Malaysian Region*, Budapest, 1982, pp. 116–17.

16 Farmer, *Agricultural Colonization*, p. 292; C. H. Gotsch and W. M. Dyer, 'Rhetoric and reason in the Egyptian New Lands debate', *Food Research Institute Studies*, 18, 1979, pp. 129–49; Goering, *Agricultural Land Settlement*.

17 FAO, *The State of Food and Agriculture 1978*, Rome, 1979, pp. 2–32.

18 Fox, 'Mexico', p. 41.

19 J. H. Galloway, 'Brazil', in Blakemore and Smith, *Latin America*, pp. 358–68; S. Cunningham, 'Recent development in the Centre-West region', *Bank of London and South America Review*, 14, 1980, pp. 44–52; R. Andrew Nickson, 'Brazilian colonization of the eastern border region of Paraguay', *Journal of Latin America Studies*, 13, 1981, pp. 111–37; J. D. Henshall and R. P. Momsen, Jr., *A Geography of Brazilian Development*, London, 1974; J. Foweraker, *The Struggle for Land: A Political Economy of the Pioneer Frontier in Brazil from 1930 to the Present Day*, Cambridge, 1981.

20 E. Allen, 'New settlement in the Upper Amazon basin', *Bank of London and South America Review*, 9, 1975, pp. 622–8.

21 W. Denevan, 'Development and the imminent demise of the Amazon rainforest', *Professional Geographer*, 25, 1973, pp. 130–5.

22 M. Hiracka, 'The development of Amazonia', *Geographical Review*, 72, 1982, pp. 94–8; E. F. Moran, *Developing the Amazon*, Bloomington, Ind., 1981;

E. F. Moran, 'Ecological, anthropological and agronomic research in the Amazon basin', *Latin American Research Review*, 171, 1982, pp. 3–41; C. Weil, 'Amazon update; developments since 1970', *Focus*, 33, 1983.

23 Denevan, *'The Amazon rainforest'*.

24 Moran, 'The Amazon basin'; M. J. Eden, 'Ecology and land development: the case of the Amazonian rainforest', *Transactions of the Institute of British Geographers*, 3 (new series), 1978, p. 444–63; J. Kirkby, 'Agricultural land use and the settlement of Amazonia', *Pacific Viewpoint*, 17, 1976, pp. 105–32; P. A. Sanchez and S. G. Salinas, 'Low input technology for managing oxisols and ultisols in tropical America', *Advances in Agronomy*, 34, 1981, pp. 279–406.

25 R. Bromley, 'The colonization of humid tropical areas in Ecuador', *Singapore Journal of Tropical Geography*, 2, 1981, pp. 15–26; J. M. Kirby, 'Colombian land use change and the development of the oriente', *Pacific Viewpoint*, 19, 1978, pp. 1–25; J. M. Kleinpenning, 'A further evaluation of the policy for the integration of the Amazon basin', *Tijdschrift voor Economische en Social Geografie*, 69, 1978, pp. 78–85.

26 D. B. Grigg, *The Agricultural Systems of the World: An Evolutionary Approach*, Cambridge, 1974, pp. 88, 92, 93, 96–108, 230–5.

27 R. Wade, 'India's changing strategy of irrigation development', in E. W. Coward, Jr. (ed.), *Irrigation and Agricultural Development in Asia; Perspectives from the Social Sciences*, Ithaca, NY, 1980, pp. 345–64.

28 B. H. Farmer, *Agricultural Colonization in South and South East Asia*, Hull, 1969; P. Blaikie, J. Cameron and D. Seddon, *Nepal in Crisis: Growth and Stagnation at the Periphery*, Oxford, 1980, p. 18.

29 G. H. Peiris, 'Land reform and agrarian change in Sri Lanka', *Modern Asian Studies*, 12, 1978, pp. 611–28; B. H. Farmer, *Peasant Colonization in Ceylon*, London, 1957; T. F. Rasmussen, 'Population and land utilization in the Assam Valley', *Journal of Tropical Geography*, 14, 1960, 51–76.

30 T. R. Tregear, *China: A Geographical Survey*, London, 1976, p. 108; R. Welch, H. C. Lo and C. W. Pannel, 'Mapping China's new agricultural lands', *Photogrammetric Engineering and Remote Sensing*, 45, 1979, pp. 1221–8.

31 K. R. Walker, 'China's grain production, 1978–80 and 1952–57; some basic statistics', *China Quarterly*, 86, 1981, pp. 215–47.

32 G. W. Jones, 'Population growth, empty land and economic development in Indonesia, the Philippines and Malaysia', *Kajtan Ekonomi Malaysia*, 5, 1968, pp. 1–18; P. Krinks, 'Old wine in a new bottle: land settlement and agrarian problems in the Philippines', *Journal of South East Asian Studies*, 5, 1974, pp. 1–17.

33 J. M. Hardjono, *Transmigration in Indonesia*, Kuala Lumpur, 1977; Jones, 'Population growth' G. W. Jones, 'Indonesia: the transmigration programme and development planning', in Pryor, *Migration and Development*, pp. 212–21.

34 World bank, *Accelerated Development in Sub-Saharan Africa: An Agenda for Action*, Washington, D.C., 1981; C. Christensen et al., *Food Problems and Prospects in Sub-Saharan Africa: The Decade of the 1980s*, International

Economics Division, Economic Research Service, United States Department of Agriculture, Washington, D.C., 1981.

35 Christensen et al., *Food Problems;* Beets, *Multiple Cropping.*

36 Christensen et al., *Food Problems;* FAO, *The State of Food and Agriculture 1989*, Rome, 1989, p. 72.

37 K. M. Barbour, 'The Sudan since independence', *Journal of Modern African Studies*, **18**, 1980, pp. 73–97; World Bank, *Accelerated Development*, pp. 14, 73, 76; M. B. K. Darkoh, 'Desertification in Tanzania', *Geography*, **67**, 1982, pp. 320–33; R. J. Harrison Church, 'Problems and development of the dry zone of West Africa', *Geographical Journal*, **127**, 1961, pp. 187–204; G. M. Higgins, A. H. Kassam, L. Naiken and M. M. Shah, 'Africa's agricultural potential', *Ceres*, **14**, 1981, pp. 13–21.

38 B. Floyd and M. Adinde, 'Farm settlements in Eastern Nigeria: a geographical approach', *Economic Geography*, **43**, 1967, pp. 189–230; T. E. Hilton, 'The Volta resettlement project', *Journal of Tropical Geography*, **24**, 1967, pp. 12–21; G. Kay, 'Resettlement and land use planning in Zambia: the Chipangali scheme', *Scottish Geographical Magazine*, **81**, 1965, pp. 163–77; R. Chambers, *Settlement Schemes in Tropical Africa: A Study of Organization and Development*, London, 1969.

39 G. V. Jacks and R. O. Whyte, *The Rape of the Earth*, London, 1939; W. Vogt, *Road to Survival*, London, 1949; M. Roberts, *The Estate of Man*, London, 1952; F. Osborn, *Our Plundered Planet*, London, 1948.

40 R. Best, *Land Use and Living Space*, London, 1981; R. H. Jackson, *Land Use in America*, London, 1981; M. Stocking, 'Measuring land degradation', in P. Blaikie and H. Brookfield (eds), *Land Degradation and Society*, London, 1987, pp. 1–26.

41 L. R. Brown, 'Soil and civilisation: the decline of food security', *Third World Quarterly*, **5**, 1983, pp. 103–17; FAO, *The State of Food and Agriculture 1977*, Rome, 1978, pp. 3–13; E. P. Eckholm *Losing Ground: Environmental Stress and World Food Prospects*, New York, 1976.

42 Eckholm, *Losing Ground*, pp. 41, 62, 94–5, 119–20, 167; Fox, 'Mexico' pp. 33, 39, 44; Blaikie et al., *Nepal in Crisis*, pp. 11–19; M. R. and A. K. Biswas, 'Loss of productive soil', *International Journal of Environmental Studies*, **12**, 1978, pp. 189–98.

43 Brown, 'Soil and civilisation'; United Nations, *Desertification: Its Causes and Consequences*, Oxford, 1977, p. 6; A. Kovda, 'Soil loss: an overview', *Agroecosystems*, **3**, 1977, 205–24.

7 Agricultural development in the developed countries since 1945

1 FAO, *Agricultural Adjustment in Developed Countries*, Rome, 1972, pp. 124–5; K. R. Gray, 'Soviet agricultural specialization and efficiency', *Soviet Studies*, **31**, 1979, pp. 542–8, footnote 1, p. 546.

2 D. Andrews, M. Mitchell and A. Weber, *The Development of Agriculture in Germany and the UK: Three Comparative Time Series, 1870–1975*, Wye

College, Ashford, Kent, Centre for European Agricultural Studies, Miscellaneous Studies No. 4, 1979, pp. 11, 60–2; E. P. Cunningham, 'The revolution in Irish agriculture, with particular reference to animal production', *Journal of the Royal Agricultural Society of England*, 141, 1980, pp. 88–98; A. Maris and J. de Veer, 'Dutch agriculture in the period 1950–1970 and a look ahead', *European Review of Agricultural Economics*, 1, 1973, pp. 63–78; J. P. O'Hagan, *Growth and Adjustment in National Agricultures*, London, 1978, p. 51; Central Statistical Office, *Economic Trends, Annual Supplement*, No. 7, 1982, table 84.

3   C. Potter, 'Approaching limits; farming contraction and environmental conservation in the UK', in D. Goodman and M. Redclift (eds), *The International Farm Crisis*, London, 1989, pp. 135–55.

4   L. S. Hardin, 'Thirty years of agriculture: a review of North America', *Span*, 30, 1988, pp. 98–101; FAO, *Production Yearbook, 1989*, 43, Rome, 1990.

5   C. Christians, 'Les resultats de 25 années de modernisation d'une agriculture avancée; l'exemple Belge', *Hommes et Terres du Nord*, 4, 1980, pp. 23–40; P. Lamartine Yates, *Food, Land and Manpower in Western Europe*, London, 1960, p. 118; M. J. Troughton, *Canadian Agriculture*, Budapest, 1982, p. 43.

6   F. Durgin, 'The Virgin Lands programme, 1954–60', *Soviet Studies*, 13, 1961–2, pp. 255–80; K. Wadekin, 'Soviet agriculture's dependence on the West', *Foreign Affairs*, 60, 1982, pp. 882–903.

7   W. W. Cochrane, *The Development of American Agriculture: A Historical Analysis*, Minneapolis, Minn., 1979, p. 162; E. O. Heady, 'The agriculture of the United States', *Scientific American*, 235, 1976, pp. 107–27.

8   C. M. Donald, 'Innovation in Australian agriculture', in D. B. Williams (ed.), *Agriculture in the Australian Economy*, Sydney, 1982, p. 64; D. P. Vincent, A. A. Powell and P. B. Dixon, 'Changes in the supply of agricultural products', in D. B. Williams, *Agriculture*, pp. 215–16.

9   G. Thiede, 'L'agriculture Européenne et la révolution technique', in M. Tracy and I. Hodac (eds), *Prospects for Agriculture in the European Economic Community*, Bruges, 1979, pp. 110–38; G. C. Fite, *American Farmers: The New Minority*, Bloomington, Ind., 1981, pp. 109–11.

10   Thiede, 'L'agriculture Européenne'; W. D. Rasmussen, 'A post-script: twenty five years of change in farm productivity', *Agricultural History*, 49, 1975, pp. 84–6.

11   S. S. Batie and R. G. Healey, 'The future of American agriculture', *Scientific American*, 248, 1983, pp. 27–35; Christians, 'Les resultats', pp. 23–40; O'Hagan, *Growth and Adjustment*; G. Brown, 'Agriculture in the EEC. 6: Denmark', *Span*, 23, 1980, pp. 29–31.

12   P. Lamartine Yates, *Food Production in Western Europe: An Economic Survey of Agriculture in Six Countries*, London, 1960; J. P. Johnson and D. J. Halliday, 'The development of fertiliser use in the UK since 1945', in A. H. Bunting (ed.), *Change in Agriculture*, London, 1970, pp. 265–73; Z. Griliches, 'Hybrid corn and the economics of innovation', *Science*, 132, 1960, pp. 275–80; J. G. Elliot, 'Weed control: past, present and future

– a historical perspective', in R. G. Hurd, P. V. Biscoe and C. Dennis (eds), *Opportunities for Increasing Crop Yields*, London, 1980, pp. 285–95; W. F. Raymond, 'Grassland research', in G. W. Cooke (ed.), *Agricultural Research 1931–1980; A History of the Agricultural Research Council and a Review of Development in Agricultural Science During the Last Fifty Years*, London, 1981.

13 T. J. Riggs, P. R. Harrison, N. D. Start, D. M. Miles, C. L. Morgan and M. D. Ford, 'Comparison of spring barley varieties grown in England and Wales between 1880 and 1980', *Journal of Agricultural Science*, 97, 1981, pp. 599–610; R. Riley, 'Plant breeding', in Cooke, *Agricultural Research*; R. B. Austin, 'Actual and potential yields of wheat and barley in the United Kingdom', *ADAS Quarterly Review*, 29, 1978, pp. 76–87; V. Silvey, 'The contribution of new varieties to increasing cereal yield in England and Wales', *Journal of the National Institute of Agricultural Botany*, 14, 1978, pp. 367–84; N. F. Jensen, 'Limits to growth in world food production', *Science*, 201, 1978, pp. 317–20.

14 I. R. Bowler, 'The agricultural pattern', in R. J. Johnston and J. C. Dornkamp (eds), *The Changing Geography of the United Kingdom*, London, 1982, pp. 75–104; A. H. Dawson, 'The great increase in barley growing in Scotland', *Geography*, 65, 1980, pp. 213–17; J. D. Palmer, 'Plant breeding today', *Journal of the Royal Agricultural Society of England and Wales*, 131, 1970, pp. 7–17; J. F. Shepherd, 'The development of new wheat varieties in the Pacific north west', *Agricultural History*, 54, 1980, pp. 52–63; G. Doussinault, A. Berbigier and M. Pollacksek, 'Trends in cereal breeding in France', *Outlook on Agriculture*, 7, 1973, pp. 222–6.

15 P. W. Russell Eggitt, 'Choosing between crops; aspects that effect the user', *Philosophical Transactions of the Royal Society of London*, series B, 28, 1977, pp. 93–106; Riley, 'Plant breeding'; K. Dexter, 'The impact of technology on the political economy of agriculture', *Journal of Agricultural Economics*, 28, 1977, pp. 211–19.

16 Heady, 'The agriculture of the United States'; Troughton, *Canadian Agriculture*, pp. 211–19; Thiede, 'L'agriculture Européenne'.

17 'European agriculture towards the end of the 20th century', *Economic Bulletin for Europe*, 25, 1983, pp. 164–5, 175; Christians, 'Les resultats', p. 40; U. Varjo, *Finnish Farming: Typology and Economics*, Budapest, 1977; P. J. Gersmehl, 'No-till farming: the regional applicability of a revolutionary agricultural technology', *Geographical Review*, 68, 1978, pp. 66–75; A. H. Kampp, *An Agricultural Geography of Denmark*, Budapest, 1975, p. 73.

18 D. W. Robinson, 'The impact of herbicides on crop production', in Hurd, Biscoe and Dennis, *Opportunities*, pp. 297–312; W. Graham-Bryce, 'Crop protection: a consideration of the effectiveness and disadvantages of current methods and scope for improvement', *Philosophical Transactions of the Royal Society of London*, series B, 281, 1977, pp. 163–79; Batie and Healey, 'American Agriculture', pp. 27–35; E. R. Bullen, 'How much cultivation?', *Philosophical Transactions of the Royal Society of London*, series B, 281, 1977, pp. 83–92.

19 F. H. W. Green, 'Recent changes in land use and treatment', *Geographical*

Journal, **142**, 1976, pp. 12–26; Batie and Healey, 'American Agriculture', pp. 27–35.

20 Wadekin, 'Soviet agriculture'; Jean Chombert de Lauve, *L'Aventure Agricole de la France de 1945 à nos jours*, Paris, 1979, p. 41; P. H. Knudsen, *Agriculture in Denmark*, Agricultural Research Council of Denmark, Copenhagen, 1977, p. 32; Kampp, *Denmark*, p. 74; J. B. Viallion, 'Croissance agricole en France et en Bourgogne de 1852–1970', *Revue d'Histoire Economique et Sociale*, **55**, 1977, pp. 464–98; E. J. Ojala, *Agriculture and Economic Progress*, London, 1952.

21 Donald, 'Australian agriculture', p. 77; Knudsen, *Agriculture in Denmark*, p. 16; Cunningham, 'Irish agriculture'.

22 Raymond, 'Grassland research'; T. H. Davies, 'The evolution of modern dairy cow grazing systems', *ADAS Quarterly Review*, **22**, 1976, pp. 275–82; D. A. Gillmore, 'Agriculture', in D. A. Gillmore (ed.), *Irish Resources and Land Use*, Dublin, 1979, p. 126.

23 J. P. Berlan, J. P. Bertrand and L. Lebas, 'The growth of the American soybean complex', *European Review of Agricultural Economics*, **4**, 1977, pp. 395–416; Knudsen, *Agriculture in Denmark*, pp. 27, 30, 37; R. E. H. Mellor, *The Two Germanies; A Modern Geography*, London, 1978, p. 291; J. T. Coppock, 'Agricultural changes in Britain', *Geography*, **49**, 1964, pp. 322–7; FAO, *Agricultural Adjustment*, pp. 50–2; J. T. Pierce, *The Food Resource*, New York, 1990, p. 39.

24 Knudsen, *Agriculture in Denmark*, p. 34; K. N. Burns, 'Diseases of farm animals,' in Cooke, *Agricultural Research*, pp. 255–76; A. S. Foot, 'Changes in milk production, 1930–1970', *Journal of the Royal Agricultural Society of England*, **131**, 1971, pp. 30–42; J. W. B. King, 'Animal breeding research in Britain 1931–1981', in Cooke, *Agricultural Research*, pp. 277–88; A. C. L. Brown, 'Animal health: present and future', *Philosophical Transactions of The Royal Society of London*, series B, **281**, 1977, pp. 181–91; Sir Keith Baxter, 'Animal nutrition', in Cooke, *Agricultural Research*, pp. 247–54.

25 Brown, 'Animal health'.

26 D. B. Bellis, 'Pig farming in the United Kingdom – its development and future trends', *Journal of the Royal agricultural Society of England*, **129**, 1968, pp. 24–42; Fite, *American Farmers*, p. 128.

8 Tropical Africa

1 FAO, *The State of Food and Agriculture 1958*, Rome, 1958, p. 97.

2 FAO, *The State of Food and Agriculture 1978*, Rome, 1979, pp. 2–3; E. H. Hartmans, 'African food production: research against time', *Outlook on Agriculture*, **12** (4), 1983, pp. 165–71; J. Borton and E. Clay, 'The African food crisis of 1982–6; a provisional view', in D. Rimmer (ed.), *Rural Transformation in Tropical Africa*, London, 1988, pp. 140–67.

3 But see estimates in C. Christensen et al., *Food Problems and Prospects in Sub-Saharan Africa: The Decade of the 1980s*, International Economics

Division, Economic Research Service, United States Department of Agriculture, Washington, D.C., 1981.

4   C. Harvey, 'The economy of sub-Saharan Africa: a critique of the World Bank's report', *African Contemporary Record: Annual Survey and Documents*, 14 (2), 1981–2, pp. A114–A119; FAO, *The State of Food and Agriculture 1989*, Rome, 1989.

5   A. T. Grove, *Africa*, London, 1978, p. 61.

6   W. T. Morgan, *East Africa*, London, 1973, pp. 130–3.

7   R. W. Steel, 'Problems of population pressure in tropical Africa', *Transactions of the Institute of British Geographers*, 49, 1970, pp. 1–14.

8   G. M. Higgins, A. H. Kassam, L. Maiken and M. M. Shah, 'Africa's agricultural potential', *Ceres*, 14 (5), 1981, pp. 13–21; see also G. M. Higgins, A. H. Kassam, L. Maiken, G. Fischer and M. M. Shah, *Potential Population Supporting Capacities of Lands in the Developing World*, Rome, 1982.

9   Higgins et al., 'Africa's agricultural potential'.

10   P. Burnham, 'Changing agricultural and pastoral ecologies in the West African Savanna region', in D. R. Harris (ed.), *Human Ecology in Savanna Environments*, London, 1980, pp. 148–50.

11   Grove, *Africa* p. 16.

12   Higgins et al., 'Africa's agricultural potential'; S. Gregory, 'Rainfall reliability', in M. F. Thomas and G. W. Whittington (eds), *Environment and Land Use in Africa*, London, 1969, pp. 57–82.

13   Grove, *Africa*, pp. 7, 11.

14   Hartmans, 'African food production'.

15   G. P. Murdock, 'Staple subsistence crops in Africa', *Geographical Review*, 50, 1960, pp. 523–40.

16   C. K. Eicher, 'Facing up to Africa's food crisis', *Foreign Affairs*, 61 (1), 1982, pp. 151–74; A. Blair Rains, 'African pastoralism', *Outlook on Agriculture*, 11 (3), 1982, pp. 96–103.

17   Lord Hailey, *An African Survey: A Study of Problems Arising in Africa South of the Sahara*, London, 1938, p. 879.

18   W. Allan, *The African Husbandman*, Edinburgh, 1955; P. H. Nye and J. D. Greenland, *The Soil under Shifting Cultivation*, Technical Communication No. 51, Commonwealth Agricultural Bureau, Harpenden, 1960.

19   R. Tourte and J. C. Moomaw, 'Traditional African systems of agriculture and their improvement', in C. L. A. Leakey and J. B. Wills (eds), *Food Crops of the Lowland Tropics*, Oxford, 1977, pp. 195–311.

20   E. Boserup, *The Conditions of Agricultural Growth*, London, 1965.

21   M. B. Gleave and H. P. White, 'Population density and agricultural systems in West Africa', in Thomas and Whittington, *Environment*, pp. 273–300.

22   J. M. Hunter and G. K. Ntiri, 'Speculations on the future of shifting agriculture in Africa', *Journal of the Developing Areas*, 12, 1978, pp. 183–208.

23   J. Heyer, J. K. Maitha and W. M. Senga, *Agricultural Development in Kenya: An Economic Assessment*, Nairobi, 1976, p. 198; W. B. Morgan, 'Peasant agriculture in tropical Africa', in Thomas and Whittington, *Environment*, pp. 241–73; M. J. Mortimore, 'Land and population pressure in

the Kano close settled zone, northern Nigeria', *Advancement of Science*, **23**, 1967, pp. 677–86.

24 Eicher, 'Africa's food crisis'; C. T. Agnew, 'Water availability and the development of rainfed agriculture in south-west Niger, West Africa', *Transactions of the Institute of British Geographers*, 7, 1982, pp. 419–57; Christensen et al., *Food Problems*; U. Lele, 'Rural Africa: modernization, equity and long term development', *Science*, **211**, 1981, pp. 547–53; R. M. Lawson, *The Changing Economy of the Lower Volta, 1954–67: A Study in the Dynamics of Rural Economic Growth*, London, 1972, pp. 40–4; P. Roberts, 'Rural development and the rural economy in Niger, 1900–75', in J. Heyer, P. Roberts and G. Williams (eds), *Rural Development in Tropical Africa*, London, 1981, pp. 1–15.

25 G. Hyden, *Beyond Ujamaa in Tanzania: Underdevelopment and an Uncaptured Peasantry*, 1980, p. 10; A Getahrun, 'Agricultural systems in Ethiopia', *Agricultural systems*, 3, 1975, pp. 281–93; B. H. Kinsey and I. Ahmed, 'Mechanical innovations on small African farms: problems of development and diffusion', *International Labour Review*, **122** (2), 1983, p. 222; Mortimore, 'Land and population pressure'; W. B. Morgan, 'Farming practice, settlement patterns and population density in south-eastern Nigeria', *Geographical Journal*, **121**, 1955, pp. 320–33.

26 K. R. M. Anthony et al., *Agricultural Change in Tropical Africa*, London, 1979, pp. 33, 88, 138; J. H. Cleave, *African Farmers: Labour Use in the Development of Smallholder Agriculture*, London, 1974, p. 16; V. Jamal, 'Getting the crisis right: missing perspectives on Africa', *International Labour Review*, **127**, 1988, pp. 655–78.

27 J. Levi and M. Havinden, *Economics of African Agriculture*, London, 1982, pp. 32–43; Anthony et al., *Agricultural Change*, pp. 38–42.

28 G. K. Helleiner, *Peasant Agriculture, Government and Economic Growth in Nigeria*, Homewood, Ill., 1966, p. 5.

29 FAO, *Food and Agriculture 1958*, p. 59.

30 FAO, *African Survey*, Rome, 1962, p. 34.

31 B. Beckman, 'Ghana, 1951–78: the agrarian basis of the post-colonial state', in Heyer, Roberts and Williams, *Rural Development*, pp. 143–67; R. A. Joseph, 'Affluence and under-development: the Nigerian experience', *Journal of Modern African Studies*, **16** (2), 1978, pp. 221–39; Anthony et al., *Agricultural Change*, pp. 38–57; FAO, *Food and Agriculture 1989*; M. F. Lofchie, 'Africa's agricultural crisis: an overview', in S. K. Commins, M. F. Lofchie and R. Payne (eds), *Africa's Agrarian Crisis: the Roots of Famine*, Boulder, Colo., 1986, pp. 3–18.

32 R. Dumont, *False Start in Africa*, London, 1966, p. 115.

33 FAO, *Production Yearbook 1980*, vol. 34, Rome, 1981.

34 Anthony et al., *Agricultural Change*, p. 37; Christensen et al., *Food Problems*; Lele, 'Rural Africa'; World Bank, *Accelerated Development in sub-Saharan Africa: An agenda for Action*, Washington, D.C., 1981, p. 47; J. M. Cohen, 'Land tenure and rural development in Africa', in R. H. Bates and M. F. Lofchie (eds), *Agricultural Development in Africa: Issues of Public Policy*, New York, 1980, pp. 349–400; B. Aklilu, 'The diffusion of fertilizer in

Ethiopia: patterns, determinants and implications', *Journal of Developing Areas*, 14, 1980, pp. 387-94; A. T. Grove, 'Geographical introduction to the Sahel', *Geographical Journal*, 144, 1978, pp. 407-15; G. B. Masefield, 'Agricultural change in Uganda, 1945-1960', *Food Research Institute Studies*, 3 (2), 1962; L. A. Paulino, 'The evolving food situation', in J. W. Mellor, C. L. Delgado and M. J. Blackie (eds), *Accelerating Food Production in sub-Saharan Africa*, Baltimore, Md., 1987, pp. 23-38; World Bank, *World Development Report 1992*, Oxford, 1993, p. 135.

35 Agnew, 'Water availability'; Roberts, 'Rural development'.

36 A. H. Kassam, M. Dagg, J. M. Kowal and F. H. Khadr 'Improving food crop production in the Sudan savanna zone of Northern Nigeria', *Outlook on Agriculture*, 8 (6), 1976, pp. 341-7.

37 W. V. Blewett, 'The farming picture in tropical Africa', *World Crops*, 2, 1950.

38 P. Wyeth, 'Economic development in Kenyan agriculture', in Tony Killick (ed.), *Papers on the Kenyan Economy: Performance, Problems and Policies*, Nairobi, 1983, pp. 299-310.

39 Anthony et al., *Agricultural Change*, p. 276; A. C. Coulson, 'Tanzania's fertilizer factory', *Journal of Modern African Studies*, 15, 1977, pp. 119-25.

40 R. E. Clute, 'The role of agriculture in African development', *African Studies Review*, 25 (4), 1982, pp. 1-20; Christensen et al., *Food Problems*.

41 U. Lele, *The Design of Rural Development: Lessons from Africa*, London, 1975, p. 46; Wyeth, 'Kenyan agriculture'; C. Delgado, 'Setting priorities for promoting African food production', in R. Cohen (ed.), *Satisfying Africa's Food Needs: Food Production and Commercilization in African Agriculture*, Boulder, Colo., 1988, pp. 1-30.

42 K. R. M. Anthony and V. C. Uchendu, 'Agricultural change in Mazabuku district, Zambia', *Food Research Institute Studies*, 9, 1970, pp. 215-67; Eicher, 'Africa's food crisis'; B. F. Johnston, 'Agricultural production potentials and small farmer strategies in sub-Saharan Africa', in Bates and Lofchie, *Agricultural Development*, pp. 67-97; M. Rukuni and C. Eicher, 'The food security equation in Southern Africa', in C. Bryant (ed), *Poverty, Policy and Food Security in Southern Africa*, Boulder, Colo., 1988, pp. 133-57.

43 World Bank, *Development Report 1992*, p. 76; M. G. Adams and J. Howell, 'Developing the traditional sector in the Sudan', *Economic Development and Cultural Change*, 27, 1979, pp. 505-18; Christensen et al., *Food Problems*, 1981, p. 29; A. M. O'Connor, *The Geography of Tropical African Development*, London, 1978; J. P. Platteau, *The Food Crisis in Africa; a Comparative Structural Analysis*, Wider Working Papers No. 44, World Institute for Development Economics Research, United Nations University, Helskinki, no date; J. Olivares, 'The agricultural development of sub-Saharan Africa: the role and potential of irrigation', *Natural Resources Forum*, 13, 1989, pp. 268-274.

44 Anthony et al., *Agricultural Change*, pp. 140, 271; L. H. Brown, 'Agricultural change in Kenya, 1945-1960', *Food Research Institute Studies*, 8, 1968, pp. 33-90; D. J. Dodge, *Agricultural Policy and Performance in Zambia; History, Prospects and Proposals for Change*, Institute of International Studies, Berkeley Research Series No. 32, 1977, p. 8.

45 A. Shepherd, 'Agrarian change in northern Ghana: public investment, capitalist farming and famine', in Heyer, Roberts and Williams, *Rural Development*; C. Uzureau, 'Animal draught in West Africa', *World Crops*, 26, 1974, pp. 112–14; Y. Orev, 'Animal draught in Africa', *World Crops*, 24, 1972, pp. 236–7; W. K. Jaeger and P. J. Matton, 'Utilization, profitibility and the adoption of animal draught power in West Africa', *American Journal of Agricultural Economics*, 72, 1990, pp. 35–48.

46 Brown, 'Kenya'.

47 Lele, *The Design of Rural Development*, p. 26.

48 Cleave, *African Farmers*.

49 Burnham, 'Changing ecologies'; J. Tosh, 'The cash-crop revolution in tropical Africa: an agricultural re-appraisal', *African Affairs*, 79, 1980, pp. 79–94.

50 Uzureau, 'Animal draught'.

51 K. C. Sharma, D. S. Prasada Rao and W. F. Shepherd, 'Productivity of agricultural labour and land: an international comparison', *Agricultural Economics*, 4, 1990, pp. 1–12.

52 V. Jamal, 'Nomads and farmers: incomes and poverty in rural Somalia', in D. Ghai and S. Radwan (eds), *Agrarian Policies and Rural Poverty in Africa*, International Labour Office, Geneva, 1983, pp. 281–311; W. Deshler, 'Livestock trypanosomiasis and human settlement in north eastern Uganda', *Geographical Review*, 50, 1960, pp. 541–54; R. Stewart, 'Prospects for livestock production in tsetse-infested Africa', *Impact of Science on Society*, 142, 1986, pp. 117–25.

53 Rains, 'African pastoralism', pp. 96–103; Paulino, 'The evolving food situation'.

54 C. Stein and C. Schultze, 'Land use and development potential in the arid regions of Kenya', *Applied Sciences and Development*, 12, 1978, pp. 47–64.

55 Heyer, Masthia and Senga, *Agricultural Development*, p. 190.

56 D. E. Vermeer, 'Collision of climate, cattle and culture in Mauritania during the 1970s', *Geographical Review*, 71, 1981, pp. 281–97; C. Colclough and P. Fallon, 'Rural poverty in Botswana: dimensions, causes and constraints', in Ghai and Radwan, *Agrarian Policies*, pp. 129–54.

57 Jamal, 'Nomads and farmers'; Stein and Schultze, 'The arid regions of Kenya'.

58 J. Bongaarts, O. Frank and R. Lesthaeghe, 'The proximate determinants of fertility in sub-Saharan Africa', *Population and Development Review*, 10, 1984, pp. 511–37; World Bank, *Population Growth and Policies in sub-Saharan Africa*, Washington, D.C., 1986, pp. 8, 12.

59 D. Anderson, *The Economics of Afforestation: a Case Study in Africa*, London, 1987; M. Stocking, 'Measuring land degradation', in P. Blackie and H. Brookfield, *Land Degradation and Society*, London, 1987, pp. 1–26; L. R. Brown and E. C. Wolf, *Soil Erosion: Quiet Crisis in the World Economy*, Washington, D.C., 1984, pp. 9, 13; R. Whitlow, 'Man's impact on vegetation: the Africa experience', in K. J. Gregory and D. E. Walling (eds), *Human Activity and Environmental Processes*, Chichester, 1987, pp. 353–80; R. A. Sedjo and M. Clawson, 'Global forests', in J. L. Simon and H. Kahn (eds), *The Resourceful Earth*, Oxford, 1984, pp. 124–70.

60 D. J. Campbell, 'The prospect for desertification in Kajiado district, Kenya', *Geographical Journal*, 152, 1986, pp. 44–55; Whitlow, 'Man's impact on vegetation'.

61 M. Hulme, 'Is environmental degradation causing drought in the Sahel. An assessment from recent empirical research', *Geography*, 74, 1989, pp. 38–46; A. Binns, 'Is desertification a myth?', *Geography*, 75, 1990, pp. 106–13; M. J. Mortimore, 'Desertification and resilience in semi-arid West Africa', *Geography*, 73, 1988, pp. 61–4.

62 Lele, *The Design of Rural Development*; D. Ghai and S. Radwan, 'Agrarian change, differentiation and rural poverty in Africa: a general survey', in Ghai and Radwan, *Agrarian Policies*, pp. 1–29; B. Dunham and C. Hines, *Agribusiness in Africa*, London, 1983, p. 120; Dodge, *Agricultural Policy*, p. 51; J. Levi, 'African agriculture misunderstood; policy in Sierra Leone', *Food Research Institute Studies*, 13, 1974, pp. 239–62; Wyeth, 'Kenyan agriculture'; J. Hinderink and J. J. Sterkenburg, 'Agricultural policy and production in Africa; the aims, the methods and the means', *Journal of Modern African Studies*, 21, 1983, pp. 1–23; R. E. Clute, 'The African food crisis: the need for reform', *Journal of Developing Societies*, 13, 1987, pp. 156–73; J. Madeley, 'The success of Cameroon's agricultural policy', *Food Policy*, 12, 1987, pp. 195–8.

63 B. N. Floyd, 'Agricultural planning in Nigeria', *Geography*, 67, 1982, pp. 345–8; T. Forrest, 'Agricultural policies in Nigeria 1900–78', in Heyer, Roberts and Williams, *Rural Development*, pp. 222–58; W. Smith, 'Crisis and response: agricultural development and Zambia's third National Development Plan', *Geography*, 66, 1981, pp. 134–6.

64 Johnston, 'Agricultural production potentials'; M. Roemer, 'Economic development in Africa: performance since independence and a strategy for the future', *Daedalus*, 111, 1982, pp. 125–48.

65 Levi and Havinden, *African Agriculture*; Hinderink and Sterkenburg, 'Agricultural Policy'; R. M. Hecht, 'The Ivory Coast economic "Miracle": what benefits for peasant farmers?', *Journal of Modern African Studies*, 21 (1), 1983, pp. 25–53; Clute, 'The African food crisis'; World Bank, *World Development Report 1986*, Washington, D.C., 1986, p. 109; *World Development Report 1989*, Washington, D.C., 1989, p. 13; *Accelerated Development in Sub-Saharan Africa: an Agenda for Action*, Washington, D.C., 1981.

66 Lele, 'Rural Africa'; M. Lipton, 'African agricultural development: the EEC's new role,' *Development Policy Review*, 1, 1983, pp. 1–21.

67 Johnston, 'Agricultural production potentials'; Harvey, 'The economy of sub-Saharan Africa'; World Bank, *Accelerated Development*, C. L. Delgado, J. W. Mellor and M. J. Blackie, 'Strategic issues in food production in sub-Saharan Africa', in Mellor, Delgado and Blackie, *Accelerating Food Production*, Md., pp. 3–22; C. Christensen, 'Food security in sub-Saharan Africa', in W. L. Hollist and F. L. Tullis (eds), *Pursuing Food Security; Strategies and Obstacles in Africa, Latin America and the Middle East*, London, 1987, pp. 67–97.

68 Anthony et al., *Agricultural Change*, pp. 249–52.

69 Lord Hailey, *An African Survey*.

70 Hartmans, 'African food production'.

71   Lele, *The Design of Rural Development*, pp. 67, 76; Anthony et al., p. 234; Clute, 'The African food crisis', P. J. Matlon, 'The West African semi-arid tropics', in Mellor, Delgado and Blackie, *Accelerating Food Production*, pp. 59–77.

72   Dunham and Hines, *Agribusiness in Africa*, p. 120.

73   C. Leys, 'African economic development in theory and practice', *Daedalus*, 111, 1982, pp. 99–124; I. L. Griffiths, 'Famine and war in Africa', *Geography*, 73, 1988, pp. 59–61.

74   Hecht, 'The Ivory Coast'; Anthony et al., *Agricultural Change*, p. 291; J. W. Bruce, 'A perspective on indigenous land tenure systems and land concentration', in R. E. Downs and S. P. Reyna (eds), *Land and Society in Contemporary Africa*, London, 1988, pp. 23–52.

75   J. Carlsen, *Economic and Social Transformation in Rural Kenya*, Scandinavian Institute of African Studies, Uppsala, 1980, p. 192; Ghai and Radwan, *Agrarian Policies*, pp. 1–29; C. L. Delgado, 'Setting priorities for promoting African food production', in R. Cohen (ed.), *Satisfying Africa's Food Needs: Food Production and Commercialization in African Agriculture*, Boulder, Colo., 1988, pp. 1–30; G. Hyden, 'Beyond hunger in Africa: breaking the spell of monoculture', in Cohen, *Satisfying Africa's Food Needs*, pp. 47–78; I. Jazcury, 'How to make Africa self sufficient in food', *Development*, 2/3, 1987, pp. 50–6.

76   J. Sadie, 'The social anthropology of economic underdevelopment', *Economic Journal*, 70, 1960, pp. 294–303; W. O. Jones, 'Economic man in Africa', *Food Research Institute Studies*, 1, 1960, pp. 107–34.

77   S. Amin, 'Underdevelopment and dependence in Black Africa – origins and contemporary forms', *Journal of Modern African Studies*, 10, 1972, pp. 503–24; M. F. Lofchie and S. K. Commins, 'Food deficits and agricultural policies in tropical Africa', *Journal of Modern African Studies*, 20, 1982, pp. 1–25,

78   FAO, *Food and Agriculture 1978*, pp. 2–5; S. Maxwell and A. Fernando, 'Cash crops in developing countries: the issues, the facts, the policies', *World Development*, 17, 1989, pp. 1677–708.

9   Latin America

1   H. Blakemore and C. T. Smith, 'Introduction', in H. Blakemore and C. T. Smith (eds), *Latin America: Geographical Perspectives*, London, 1983, p. 12; C. T. Smith, 'The central Andes', in Blakemore and Smith, *Latin America*, pp. 278, 295.

2   P. Lamartine Yates, *Mexico's Agricultural Dilemma*, Tucson, Ariz., 1981, pp. 41, 60; FAO, *Prospects for Agricultural Development in Latin America*, Rome, 1954, p. 69; Inter-American Development Bank, *Economic and Social Progress in Latin America; Natural Resources*, Washington, D.C., 1982, p. 21; FAO, *Provisional Indicative World Plan for Agricultural Development*, vol. 1, Rome, 1970, p. 44.

3   'Agriculture in Latin America: problems and prospects', *Economic Bulletin*

of Latin America, **8**, 1963, pp. 147–94; N. Gligo, 'The environmental dimension in agricultural development', Cepal Review, **12**, 1980, pp. 129–43; World Bank, World Development Report 1989, Washington, D.C., 1989.

4   Alan Gilbert, Latin America, London, 1990, p. 27; FAO, The State of Food and Agriculture 1989, Rome, 1989, p. 46.

5   E. V. Iglesias, 'The ambivalence of Latin America agriculture', Cepal Review, **6**, 1978, pp. 7–18; F. C. Turner, 'The rush to the cities in Latin America', Science, **192**, 1976, pp. 955–62.

6   D. A. Preston, 'Rural emigration and the future of agriculture in Ecuador', in D. A. Preston (ed.), Environment, Society and Rural Change in Latin America, Chichester, 1980, pp. 195–208; L. S. Williams and E. C. Griffin, 'Rural and small town depopulation in Colombia', Geographical Review, **68**, 1978, pp. 13–30; Smith, 'The central Andes', pp. 279–90.

7   A. Berry, 'Agrarian structure, rural labour markets and trends in rural incomes in Latin America', in V. L. Urquidi and S. T. Reyes (eds), Human Resources, Employment and Development, vol. 4: Latin America, London, 1983, p. 180; M. S. Grindle, State and Countryside. Development Policy and Agrarian Politics in Latin America, Baltimore, Md., 1986, p. 98.

8   P. Dorner and R. Quiros, 'Institutional dualism in Central America's agricultural development', Journal of Latin American Studies, **5**, 1973, pp. 217–32; E. Feder, The Rape of the Peasantry; Latin America's Landholding System, New York, 1971, pp. 29, 33; S. Barraclough, 'Rural development and employment prospects in Latin America', in A. J. Field (ed.), City and Country in the Third World: Issues in the Modernization of Latin America, Cambridge, Mass., 1970, pp. 97–135; G. Gomez and A. Perez, 'The process of modernization in Latin American agriculture', Cepal Review, **8**, 1979, pp. 55–74; M. J. Smarkis and T. Saravi, Reactivating Agriculture: Strategy for Development, Ninth Inter-American Conference of Ministers of Agriculture, San Jose, Costa Rica, 1987, p. 39.

9   S. Barraclough, Agrarian Structure in Latin America, Lexington, Mass., 1973.

10   Feder, The Rape of the Peasantry, pp. 17–18; Alain de Janvry, The Agrarian Question and Reformism in Latin America, Baltimore, Md., 1981, p. 111; S. Barraclough and A. L. Domike, 'Agrarian structure in seven Latin American countries', Land Economics, **42**, 1960, pp. 391–424.

11   De Janvry, The Agrarian Question, p. 131; Feder, The Rape of the Peasantry, 54; P. Harrison, 'The inequities that curb potential', Ceres, **8**, 1981, pp. 22–6; P. Peek, Agrarian Change and Rural Emigration in Latin America, International Labour Organization, Geneva, 1978.

12   D. J. Fox, 'Mexico', in Blakemore and Smith, Latin America, p. 47; C. T. Smith, 'Land reform as a precondition for Green Revolution in Latin America', in T. P. Bayliss-Smith and Sudhir Wanmali (eds), Understanding Green Revolutions: Agrarian Change and Development Planning in South Asia, Cambridge, 1984, pp. 18–36; C. Kay, 'Achievements and contradictions of the Peruvian agrarian reform', Journal of Development Studies, **18**, 1981, pp. 141–70; C. S. Blankstein and C. Zuvekas, Jr., 'Agrarian reform in Ecuador: an evaluation of past efforts and the development of a new approach', Economic Development and Cultural Change, **22**, 1973, pp.

73–94; D. Browning, 'Agrarian reform in El Salvador', *Journal of Latin American Studies*, 15, 1983, pp. 399–426.

13 Smith, 'The central Andes', 1983, p. 292; V. Bulmer-Thomas, 'Economic development in the long run – Central America since 1920', *Journal of Latin American Studies*, 15, 1983, pp. 269–94; de Janvry, *The Agrarian Question*, pp. 121–2.

14 M. M. Cole, '*Cerrado, caatinga and pantanal*; the distribution and origin of the savanna vegetation of Brazil', *Geographical Journal*, 126, 1960, pp. 168–79.

15 R. C. West and J. P. Augelli, *Middle America: Its Lands and People*, New York, 1966, pp. 336–7.

16 A. S. Morris, *South America*, London, 1979, pp. 27–29.

17 Gomez and Perez, 'The process of modernization'; Yates, *Mexico's Agricultural Dilemma*, p. 60.

18 C. Dozier, 'Agriculture and development in Mexico's Tabasco lowlands: planning and potential', *Journal of Developing Areas*, 5, 1970, pp. 61–72; J. Revel-Mouroz, 'Mexican colonization experience in the humid tropics', in Preston, *Environment*, pp. 83–102; Yates, *Mexico's Agricultural Dilemma*, p. 47; C. Weil, 'Migration among landholdings by Bolivian campesinos', *Geographical Review*, 73, 1983, pp. 182–97; A. Delavaud, 'From colonization to agricultural development: the case of coastal Ecuador', in Preston, *Environment*, pp. 67–81; M. Nelson, *The Development of Tropical Lands: Policy Issues in Latin America*, Baltimore, Md., 1973, p. 22.

19 J. H. Galloway, 'Brazil', in Blakemore and Smith, *Latin America*, p. 359; D. E. Goodman, 'Rural structure, surplus mobilisation, and modes of production in a peripheral region: the Brazilian north-east', *Journal of Peasant Studies*, 5, 1977, pp. 3–32; B. Bret, 'Données et réflexions sur l'agriculture Brésilienne', *Annales de Geographié*, 84, 1975, pp. 557–88.

20 FAO, *Prospects for Agricultural Development*; 'Agriculture in Latin America: problems and prospects', *Economic Bulletin of Latin America*, 8, 1963, pp. 147–94; J. D. Henshall and R. P. Momsen, Jr., *A Geography of Brazilian Development*, London, 1974, p. 76.

21 Gomez and Perez, 'The process of modernization', p. 46; C. Hewitt de Alcantara, *Modernizing Mexican Agriculture: Socioeconomic Implications of Technological Changes, 1940–70*, Geneva, 1976; C. Hewitt de Alcantara, 'The Green Revolution as history: the Mexican experience', *Development and Change*, 5, 1973–4, pp. 25–44; L. Harlan Davis, 'Foreign aid to the small farmer: the El Salvador experience', *Latin American Economic Affairs*, 29, 1975, pp. 81–91; FAO, *Production Yearbook 1981*, vol. 35, Rome, 1982; M. Lipton and R. Longhurst, *New Seeds and Poor People*, London, 1989, p. 2.

22 D. G. Dalyrmple, *Development and Spread of High Yielding Varieties of Wheat and Rice in the Less Developed Nations*, Washington, D.C., 1978; G. M. Scobie and R. Posada, 'The impact of technical change on income distribution: the case of rice in Columbia', *American Journal of Agricultural Economics*, 60, 1978, pp. 85–92; W. C. Thiesenhusen, 'Green Revolution in Latin America: income effects, policy decisions', *Monthly Labour*

Review, **95**, 1972, pp. 20–7; E. J. Wellhausen, 'The agriculture of Mexico', *Scientific American*, **235**, 1976, pp. 129–50.

23 FAO, *Prospects for Agricultural Development*; FAO, *Annual Fertilizer Review*, Rome, 1982, vol. 32, 1983; de Janvry, *The Agrarian Question*, p. 164.

24 E. Ortega, 'Peasant agriculture in Latin America', *Cepal Review*, **16**, 1982, pp. 75–111; R. M. Mendes, 'The rural sector in the socio-economic context of Brazil', *Cepal Review*, **33**, 1987, pp. 39–59.

25 J. H. Saunders and V. W. Ruttan, 'Biased choice of technology in Brazilian agriculture', in H. P. Binswanger and V. W. Ruttan (eds), *Induced Innovation: Technology, Institutions and Development*, Baltimore, Md., 1978, pp. 276–96; J. Saunders, 'The modernization of Brazilian society', in J. Saunders (ed.), *Modern Brazil: New Patterns and Development*, Gainesville, Fla., 1971, pp. 1–28.

26 Ortega, 'Peasant agriculture'; Bulmer-Thomas, 'Central America since 1920'; E. Ortega, 'Agriculture as viewed by ELCAC', *Cepal Review*, **35**, 1988, pp. 13–40.

27 R. Pebayle, 'Rural innovation and the organization of space in southern Brazil', in Preston, *Environment*, pp. 103–19; de Alcantara, *Modernizing Mexican Agriculture*.

28 D. W. Adams, 'Agricultural credit in Latin America', *American Journal of Agricultural Economics*, **53**, 1971, pp. 163–72; D. Goodman and M. Redclift, *From Peasant to Proletarian: Capitalist Development and Agrarian Transitions*, Oxford, pp. 145–8; de Janvry, *The Agrarian Question*, pp. 158–60; de Alcantara, 'The Green Revolution'.

29 Henshall and Momsen, *A Geography*, p. 99; FAO, *Production Yearbook 1981*, vol. 35, Rome, 1982; L. L. Cordovez, 'Trends and recent changes in the Latin American food and agriculture situation', *Cepal Review*, **16**, 1982, pp. 7–14. de Alcantara, 'The Green Revolution'; Wellhausen, 'The agriculture of Mexico'; Gomez and Perez, 'The process of modernization'; E. Boyd Wennergren and M. D. Whitaker, *The Status of Bolivian Agriculture*, New York, 1975, p. 111; de Janvry, *The Agrarian Question*, p. 160.

30 R. A. Berry, 'Land distribution, income distribution and the productive efficiency of Columbian agriculture', *Food Research Institute Studies*, **72**, 1973, pp. 199–232.

31 Ortega, 'Peasant agriculture'; de Janvry, *The Agrarian Question*, pp. 132–3, 160; L. H. Davis and D. E. Weisenhausen, 'Small farmer market development: the El Salvador experience', *Journal of Developing Areas*, **15**, 1981, pp. 407–16.

32 Ortega, 'Peasant agriculture'; Berry, 'Land distribution'.

33 C. T. Smith, 'Aspects of agriculture and settlement in Peru', *Geographical Journal*, **126**, 1960, pp. 297–412; Cordovez, 'Trends and recent changes'; W. H. Durham, *Scarcity and Survival in Central America: Ecological Origins of the Soccer War*, Stanford, 1979, pp. 30–3, 34; Dorner and Quiros, 'Institutional dualism'; Bulmer-Thomas, 'Central America since 1920'; R. Burbach and P. Flynn, *Agribusiness in the Americas*, New York, 1980, p. 105; Inter-American Development Bank, *Economic and Social Progress*, p. 119; de Janvry, *The Agrarian Question* p. 90; F. H. de Melo, 'The external

crisis adjustment policies and agricultural development in Brasil', *Cepal Review*, **33**, 1987, pp. 83–90; G. Salgado, 'Ecuador: crisis and adjustment policies: their effect on agriculture', *Cepal Review*, **33**, 1987, pp. 129–43; FAO, *Production Yearbook 1990*, vol. 44, Rome, 1991, p. 107.

10   Asia

1   A. Doak Barnett, *China and the World Food System*, London, 1979, p. 65; R. C. Y. Ng, 'Development and change in rural Thailand', *Asian Affairs*, **10**, 1979, pp. 62–8; P. Pinstrup-Andersen and Peter B. R. Hazell, 'The impact of the Green Revolution and prospects for the future', in J. P. Gittinger, J. Leslie and C. Hoisington (eds), *Food Policy: Integrating Supply, Distribution, Consumption*, Baltimore, Md., 1987, pp. 106–18.

2   D. Grigg, *An Introduction to Agricultural Geography*, 1984, pp. 186–92; P. J. Atkins, 'Operation Flood; dairy development in India', *Geography*, **74**, 1989, pp. 259–62.

3   J. G. Gurley, 'Rural development in China 1949–75, and the lessons to be learned from it', in N. Maxwell (ed.), *China's Road to Development*, London, 1979, pp. 5–26; Reeitsu Kojima, 'China's new agricultural policy', *Developing Economics*, **20**, 1982, pp. 390–413; J. Gray, 'China's new agricultural revolution', *IDS Bulletin*, **13**, 1982, pp. 36–43; D. Elz, 'From collectivisation to market orientation – a review of agricultural reform in China', *Quarterly Journal of Agriculture*, **28**, 1989, pp. 154–65; D. Gale Johnson, 'Economic reforms in the Peoples Republic of China', *Economic Development and Cultural Change*, Supplement, **36**, 1988, pp. 225–45; J. T. Pierce, *The Food Resource*, London, 1990, p. 104.

4   B. L. C. Johnson, *India: Resources and Development*, London, 1979, p. 20; 'Prospects for the economic development of Bangladesh in the 1980s', *Economic Bulletin for Asia and the Pacific*, **31**, 1980, pp. 94–105.

5   Cheng-Hung Liao and M. Yand, 'Socio-economic change in rural Taiwan, 1950–78', *South East Asian Studies*, **18**, 1981, pp. 539–45.

6   B. H. Farmer, *An Introduction to South Asia*, London, 1984, pp. 188, 210, 212; A. Bhaduri, 'A comparative study of land reform in South Asia', *Economic Bulletin for Asia and the Pacific*, **30**, 1979, pp. 1–13.

7   P. Krinks, 'Rural changes in Java: an end to involution?', *Geography*, **63**, 1978, pp. 31–6.

8   FAO, *The State of Food and Agriculture 1978*, Rome, 1979, p. 22.

9   Japan FAO Association, *A Century of Technical Development in Japanese Agriculture*, Tokyo, 1959; Sung Hwan Ban, 'Agricultural growth in Korea 1918–1971', in Y. Hayami, V. W. Ruttan and H. M. Southworth, *Agricultural Growth in Japan, Taiwan, Korea and the Philippines*, Honolulu, Hawaii, 1979, pp. 90–116; S. Pao-San Ho, 'Agricultural transformation under colonialism: the case of Taiwan', *Journal of Economic History*, **28**, 1965, pp. 313–40; D. J. Puchala and J. Stavely, 'The political economy of Taiwanese agricultural development', in R. E. Hopkins, D. J. Puchala and R. B. Talbot (eds), *Food, Politics and Agricultural Development: Case Studies*

in the Public Policy of Rural Modernization, Boulder, Colo. 1979, pp. 112–17; S. McCune, *Korea's Heritage: A Regional and Social Geography*, Tokyo, 1957, pp. 85–6; Y. Hayami, 'Elements of induced innovation: a historical perspective for the Green Revolution', *Explorations in Economic History*, **8**, 1971, pp. 445–72.

10 R. H. Myers, 'Land, property rights and agriculture in modern China', in R. Barker and R. Sinha (eds), *The Chinese Agricultural Economy*, London, 1982, pp. 37–47; G. Blyn, *Agricultural Trends in India, 1891–1947*, New York, 1966.

11 R. H. Day and Inderjit Singh, *Economic Development as an Adaptive Process: The Green Revolution in the Indian Punjab*, Cambridge, 1977, p. 51; C. C. David and R. Barker, 'Agricultural growth in the Philippines, 1948–1971', in Hayami, Ruttan and Southworth, *Agricultural Growth*, pp. 117–42; D. Feeny, *The Political Economy of productivity: Thai Agricultural Development 1880–1975*, Vancouver, 1982, p. 44.

12 A. Booth, 'Indonesian agricultural development in comparative perspective', *World Development*, **17**, 1989, pp. 1235–54; N. R. Lardy, 'Prospects and some policy problems of agricultural development in China', *American Journal of Agricultural Economics*, **68**, 1986, pp. 450–7.

13 P. Beaumont, 'Wheat production and the growing food crisis in the Middle East', *Food Policy*, **14**, 1989, pp. 378–84; V. Smil, 'Food production and quality of diet in China', *Population and Development Review*, **12**, 1986, pp. 25–45; V. Smil, 'China's food', *Scientific American*, **253**, 1985, pp. 104–12; J. S. Sarma, 'India – a drive towards self-sufficiency in food grains', *American Journal of Agricultural Economics*, **60**, 1978, pp. 859–64.

14 Sarma, 'India'; K. R. Walker, 'China's grain production 1975–80 and 1952–7; some basic statistics', *China Quarterly*, **86**, 1981, pp. 215–47.

15 P. Buringh, H. D. J. van Heemst and G. J. Staring, *Computation of the Absolute Maximum Food Production of the World*, Wageningen, 1975; President's Science Advisory Committee, *The World Food Problem*, vol. 2, Washington, D.C., 1967, p. 434.

16 J. E. Spencer, *Shifting Cultivation in Southeastern Asia*, Berkeley, Calif., 1966.

17 FAO, *The State of Food and Agriculture 1982*, Rome, 1983, p. 65; M. Lipton, *New Seeds and Poor People*, London, 1989, p. 9.

18 A. Pearse, *Seeds of Plenty, Seeds of Want: Social and Economic Implications of the Green Revolution*, Oxford, 1980; K. Griffin, *The Political Economy of Agrarian Change: An Essay on the Green Revolution: Economic Gains and Political Costs*, Princeton, N. J., 1971; B. Sen, *The Green Revolution in India: A Perspective*, New Delhi, 1974; B. H. Farmer, *Green Revolution? Technology and Change in Rice Growing Areas of Tamil Nadu and Sri Lanka*, London, 1977; M. Ghaffar Chaudhury, 'Green Revolution and rural incomes: Pakistan's experience', *Pakistan Development Review*, **21**, 1982, pp. 173–205; M. Prahladachar, 'Income distribution effects of the Green Revolution in India: a review of empirical evidence', *World Development*, **11**, 1983, pp. 927–44; D. G. Dalrymple, 'The adoption of high yielding grain varieties in developing nations', *Agricultural History*, **53**, 1979, pp. 704–26; C. E. Pray, 'The Green Revolution as a case study in transfer of technology',

Annals of the American Academy of Political and Social Science, **458**, 1981, pp. 68–80; V. W. Ruttan and H. P. Binswanger, 'Induced innovation and the Green Revolution', in H. P. Binswanger and V. W. Ruttan (eds), *Induced Innovation: Technology, Institutions and Development*, Baltimore, Md., 1978, pp. 358–408.

19   B. H. Farmer, *South Asia*, p. 177; R. J. Herring, 'The dependent welfare state: nutrition; entitlements and exchange in Sri Lanka', in W. L. Hollist and F. L. Tullis (eds), *Pursuing Food Security, Strategies and Obstacles in Africa, Asia, Latin America and the Middle East*, Boulder, Colo., 1987, pp. 158–80; Lipton, *New Seeds*, p. 4.

20   Walker, 'China's grain production'.

21   B. H. Farmer, 'The Green Revolution in South Asian ricefields: environment and production', *Journal of Development Studies*, **15**, 1978, pp. 304–19.

22   Yhi-Min Ho, *Agricultural Development of Taiwan, 1903–60*, Kingsport, Tenn. 1966, p. 87.

23   T. B. Wiens, 'Technological change', in Barker and Sinha, *The Chinese Agricultural Economy*, pp. 99–120: D. G. Dalrymple, *Development and Spread of High Yielding Varieties of Wheat and Rice in the Less Developed Nations*, United States Department of Agriculture, Washington, D.C., 1978; C. J. Baker, 'Frogs and farmers: the Green Revolution in India and its murky past', in T. P. Bayliss-Smith and S. Wanmali (eds), *Understanding Green Revolutions*, Cambridge, 1984, pp. 37–52.

24   E. Dayal, 'Regional responses to high yield varieties of rice in India', *Singapore Journal of Tropical Geography*, 4, 1983, pp. 87–98; D. Byerlee and B. Curtis, 'Wheat: a crop transformed', *Span*, 30, 1988, pp. 110–13; T. R. Hargrove, 'Rice production leaps forward', *Span*, 30, 1988, pp. 114–15; B. H. Farmer, 'Perspectives on the Green Revolution in South Asia', *Modern Asian Studies*, **20**, 1986, pp. 175–99.

25   FAO, *Food and Agriculture 1978*, p. 22.

26   R. Barker, D. G. Sisler and B. Rose, 'Prospects for growth in grain production', in R. Barker and R. Sinha (eds), *The Chinese Agricultural Economy*, London, 1982, pp. 37–47. pp. 163–82; H. J. Groen and J. A. Kilpatrick, 'Chinese agricultural production', in Joint Economic Committee, *Chinese Economy Post Mao: A Compendium of Papers Submitted to the Joint Economic Committee Congress of the United States, vol. 1: Policy and Performance*, Washington, D.C., 1978, pp. 607–52; R. M. Field and J. A. Kilpatrick, 'Chinese grain production: an interpretation of the data', *China Quarterly*, 74, 1978, pp. 369–84; R. C. Hsu, *Food for One Billion: China's Agriculture since 1949*, Boulder, Colo., 1982, p. 65.

27   M. Lipton, *New Seeds*, pp. 2, 400.

28   Y. Hayami, 'Induced innovation, Green Revolution and income distribution: comment', *Economic Development and Cultural Change*, 30, 1981, pp. 169–76; M. Prahladachar, 'Income distribution effects'; P. Flores-Moya, R. E. Evenson and Y. Hayami, 'Social returns to rice research in the Philippines: domestic benefits and foreign spillover', *Economic Development and Cultural Change*, **26**, 1978, pp. 591–607; C. P. Timmer and W. P. Falcon, 'The political economy of rice production and trade in Asia', in

L. Reynolds (ed.), *Agriculture in Development Theory*, Princeton, N.J., 1975, pp. 373–408; M. Rezaul Karim, *The Food Population Dilemma in Bangladesh*, Comilla, 1980, p. 63; J. Rigg, 'The Green Revolution and equity. Who adopts the new rice varieties and why', *Geography*, 74, 1989, pp. 144–50.

29 Kuan-I Chen and R. T. Tsuchigane, 'An assessment of China's food grain supplies in 1980', *Asian Survey*, 16, 1976, pp. 931–47; S. Ishikawa, 'Prospects for the Chinese economy in the 1980s', *Economic Bulletin for Asia and the Pacific*, 31, 1980, pp. 1–30; Gurley, 'China 1949–75'; T. B. Wiens, 'The evolution of policy and capabilities in China's agricultural technology', in Joint Economic Committee, *Chinese Economy Post Mao*, pp. 671–703; T. G. Rawski, 'Agricultural employment and technology', in Barker and Sinha, *The Chinese Agricultural Economy*, pp. 121–36; C. W. Pannell and L. J. C. Ma, *China: The Geography of Development and Modernization*, 1983, p. 130.

30 David and Barker, 'Agricultural growth', pp. 127–42; F. D. O'Reilly and P. I. McDonald, *Thailand's Agriculture*, Budapest, 1983, p. 54.

31 J. D. Gavan and J. A. Dixon, 'India; a perspective on the food situation', *Science*, 188, 1975, pp. 541–9; Hsu, *Food for One Billion*, pp. 53, 58; Wiens, 'The evolotion of policy'; Barnett, *China*, p. 50; R. W. Herdt and T. H. Wickham, 'Exploring the gap between potential and actual rice yield in the Philippines', *Food Research Institute Studies*, 14, 1975, pp. 163–81; L. R. Brown, *The Changing World Food Prospect; The Nineties and Beyond*, World Watch Paper 85, Washington, D.C., 1988, p. 31.

32 V. Smil, 'Controlling the Yellow River', *Geograpical Review*, 69, 1979, pp. 253–72; Gavan and Dixon, 'India'; R. Wade, 'India's changing strategy of irrigation development', in E. W. Coward, Jr. (ed.), *Irrigation and Agricultural Development in Asia*, Ithaca, NY, 1980, pp. 345–64; J. Gray, 'China's new agricultural revolution', *IDS Bulletin*, 13, 1982, pp. 36–43; Groen and Kilpatrick, 'Chinese agricultoral production'; B. Stone, 'The use of agricultural statistics', in Barker and Sinha, *The Chinese Agricultural Economy*, pp. 205–45; W. R. Gasse, *Survey of Irrigation in Eight Asian Nations*, United States Department of Agriculture, Washington, D.C., 1981; R. F. Dernberger, 'Agriculture in Communist development strategy', in Barker and Sinha, *The Chinese Agricultural Economy*, pp. 65–79.

33 FAO, *Food and Agriculture 1978*, p. 22.

34 Dayal, 'Regional responses'; Wiens, 'Technological change'.

35 Hsu, *Food for One Billion*, p. 82; G. Etienne, *India's Changing Rural Scene 1963–1979*, Delhi, 1982, p. 180.

36 R. H. Brannon, C. T. Alton and J. T. Davis, 'Irrigated dry season crop production in north east Thailand; a case study', *Journal of Developing Areas*, 14, 1980, pp. 191–200.

37 E. Dayal, 'Impact of irrigation expansion on multiple cropping in India', *Tijdschrift voor economische en Sociale Geografie*, 68, 1977, pp. 100–9; B. Dasgupta, *Agrarian Change and the New Technology in India*, Geneva, 1977, p. 90; F. Leeming, 'Progress towards triple cropping in China', *Asian Survey*, 19, 1979, pp. 450–67.

38 D. G. Dalrymple, *Survey of Multiple Cropping in Less Developed Nations*, US, Department of Agriculture, Washington, D.C., 1971; D. S. Gibbons, R. de Koninck and Ibrahim Hasan, *Agricultural Modernization, Poverty and Inequality: The Distributional Impact of the Green Revolution in Regions of Malaysia and Indonesia*, London, 1981, p. 7.

39 Leeming, 'Progress towards triple cropping'; Walker, 'China's grain production'; G. T. Trewartha and Shou-Jen Yang, 'Notes on rice growing in China', *Annals of the Association of American Geographers*, 38, 1948, pp. 277–81; Wu Chuan-chun, 'The transformation of the agricultural landscape in China', L. J. C. Ma and A. G. Noble (eds), *The Environment: Chinese and American Views*, London, 1981, pp. 35–43.

40 Stone, 'Agricultural statistics', pp. 210–11.

41 J. E. Nicklum, 'Labour accumulation in rural China and its role since the cultural revolution', *Cambridge Journal of Economics*, 2, 1978, pp. 273–86; S. de Vylder, *Agriculture in Chains: Bangladesh, A Study in Contradictions and Constraints*, London, 1982, p. 11; C. F. Framingham, Sommuk Sriplung and E. O. Heady, 'Agricultural situation and policy issues', in K. J. Nicol, S. Sriplung and O. Heady (eds), *Agricultural Development in Thailand*, Ames, Iowa, 1982; Lardy, 'Agricultural development in China'.

42 Hsu, *Food for One Billion*, pp. 33, 78, 130; E. B. Vermeer, 'Income differentials in rural China', *China Quarterly*, 89, 1982, pp. 1–33; Gale Johnson, 'Economic reforms'; Kuan-I Chen, 'China's food policy and population' *Current History*, 80, 1987, pp. 257–60, 274–6.

43 Asian Development Bank, *Rural Asia Challenge and Opportunity*, New York, 1977, p. 53; T. B. Wiens, 'Agriculture in the four modernizations', in C. W. Pannell and C. L. Salter (eds), *China Geographer*, vol. 11, *Agriculture*, Boulder, Colo., 1981, pp. 57–72. Ng, 'Development and change'; Krinks, 'Rural changes in Java.'

44 R. D. Hill, *Agriculture in the Malaysian Region*, Budapest, 1982, p. 100.

45 Dasgupta, *Agrarian Change*; FAO, *Production Yearbook 1981*, vol. 35, Rome, 1982, p. 274; FAO, *Production Yearbook 1990*, vol. 44, Rome, 1991.

46 Stone, 'Agricultural statistics', p. 240.

47 Wiens, 'Technological change'; Hsu, *Food for One Billion*, p. 80.

11 Trade and aid

1 P. Bairoch, 'Agriculture and the industrial revolution', in C. M. Cipolla (ed.), *The Industrial Revolution*, London, 1973, pp. 452–506; K. Campbell, *Food for the Future; How Agriculture can Meet the Challenge*, London, 1979, p. 123; S. Millman, S. Aronson, L. M. Fruzzetti, M. Hollos, R. Okello and Van Whiting, Jr. 'Organisation, information and entitlement in the emerging global food system', in L. F. Newman (ed.) *Hunger in History. Food Shortage, Poverty and Deprivation*, Oxford, 1990, pp. 307–30.

2 R. M. Stern, 'A century of food exports', *Kyklos*, 13, 1960, pp. 44–57.

3 Stern, 'A century of food exports'; FAO, *Trade Yearbook 1988*, vol. 42, Rome, 1990.

4   G. Bastin and J. Ellis, *International Trade in Grain and the World Food Economy*, The Economist Intelligence Unit, Special Report No. 83, London, 1980, p. 93.
5   FAO, *The State of Food and Agriculture 1982*, Rome, 1983, p. 82.
6   R. L. Paarlberg, 'Shifting and sharing adjustment burdens: the role of the industrial food importing nations', in R. F. Hopkins and D. J. Puchala (eds), *The Global Political Economy of Food*, Madison, Wis., 1978, pp. 79–101.
7   Bastin and Ellis, International Trade, p. 88; H. Wagstaff, 'Food imports of developing countries', *Food Policy*, 7, 1982, pp. 57–68; FAO, *Trade Yearbook 1981*, vol. 35, Rome, 1982; *Trade Yearbook 1988*, vol. 42, Rome, 1990; D. Byerlee, 'The political economy of Third World food imports: The case of wheat'; *Economic Development and Cultural Change*, 35, 1986, pp. 307–28.
8   Wagstaff, 'Food imports'; World Bank, *World Development Report 1981*, Washington, D.C., 1981, p. 102; J. W. Mellor, 'Third World development: food, employment and growth interactions', *American Journal of Agricultural Economics*, 64, 1982, pp. 304–11.
9   J. P. O'Hagan, 'National self sufficiency in food', *Food Policy*, 1, 1976, pp. 355–66; 'Self-sufficiency: facts and figures', *Ceres*, 12, 1979, pp. 19–21; S. F. Fallows and J. V. Wheelcock, 'Self sufficiency and United Kingdom food policy', *Agricultural Administration*, 11, 1982, pp. 107–25.
10 T. Jackson and D. Eade, *Against the Grain*, Oxford, 1982; J. Cathie, *The Political Economy of Food Aid*, Aldershot, 1982.
11 FAO, *The State of Food and Agriculture 1977*, Rome, 1978, p. 137.
12 J. R. Tarrant, 'The geography of food aid', *Transactions of the Institute of British Geographers*, 5, 1980, pp. 125–40; E. J. Clay, 'Is European Community food aid reformable?', *Food Policy*, 8, 1983, pp. 174–7; FAO, *The State of Food and Agriculture 1981*, Rome, 1982, pp. 1–27.
13 S. J. Maxwell and H. W. Singer, 'Food aid to developing countries: a survey', *World Development*, 7, 1979, pp. 225–46; J. R. Tarrant, 'EEC food aid', *Applied Geography*, 2, 1982, pp. 127–41; FAO, *The State of Food and Agriculture 1982*, Rome, 1983, p. 41: FAO, *Food Aid in Figures 1990*, 8 (1), Rome, 1990.
14 Cathie, *The Political Economy.*, p. 9; R. Vengroff, 'Food aid dependency: PL 480 aid to Black Africa', *Journal of Modern African Studies*, 20, 1982, pp. 17–43.
15 FAO, *The State of Food and Agriculture 1982*, Rome, 1983, p. 41; Maxwell and Singer, *'Food aid'*; P. J. Isenman and H. W. Singer, 'Food aid disincentive effects and their policy implications', *Economic Development and Cultural Change*, 25, 1977, pp. 205–37; L. Dudley and R. J. Sandilands, 'The side effects of foreign aid: the case of Public Law 480 wheat in Colombia', *Economic Development and Cultural Change*, 23, 1975, pp. 325–36.
16 Jackson and Eade, *Against the Grain;* H. Singer, J. Wood and Tony Jennings, *Food aid; the Challenge and the Opportunity*, Oxford, 1987, p. 123.
17 FAO, *Current World Food Situation*, Rome, 1989, p. 7.

18 R. Duncan and E. Lutz, 'Penetration of industrial country markets by agricultural products from developing countries', *World Development*, **11**, 1983, pp. 771–86; FAO, *Food and Agriculture 1982*, p. 18; FAO, *Trade Yearbook 1988*, vol. 42, Rome, 1990; T. Josling, 'The changing role of developing countries in international trade', in E. Clay and J. Shaw (eds), *Poverty, Development and Food: Essays in Honour of H. W. Singer*, 1987, London, pp. 42–58.

19 FAO, *The State of Food and Agriculture 1981*, Rome, 1982, pp. 61–6; FAO, *Trade Yearbook 1988*, vol. 42, Rome, 1990.

20 C. Mackel, J. Marsh and B. Revell, 'Western Europe and the South: the Common Agricultural Policy', *Third World Quarterly*, **6**, 1984, pp. 131–44; M. E. Abel, 'The impact of U.S. agricultural policies on the trade of developing countries', F. S. Tolley and P. A. Zadrozny (eds), *Trade, Agriculture and Development*, Cambridge, Mass., 1975, pp. 21–56; FAO, *The State of Food and Agriculture 1973*, Rome, 1973, p. 13; A. J. Rayner, K. A. Ingersent and R. C. Hine, 'Agriculture in the Uruguay round; prospects for long term trade reform', *Oxford Agrarian Studies*, **18**, 1990, pp. 3–21; A. H. Sarris, 'Prospects for EC agricultural trade with developing countries', *European Review of Agricultural Economics*, **16**, 1989, pp. 69–186; H. Corbet, 'Agricultural issues at the heart of the Uruguay Round', *National Westminster Quarterly Review*, August, 1991, pp. 2–19.

21 FAO, *The State of Food and Agriculture 1976*, Rome, 1977, pp. 1–40; FAO, *The State of Food and Agriculture 1960*, Rome, 1960, pp. 46–7.

12 Conclusions

1 D. Arnold, *Famine, Social Crisis and Historical Change*, Oxford, 1988, p. 20; W. A. Dando, *The Geography of Famine*, London, 1980.

2 M. Aymard, 'Toward the history of nutrition: some methodological remarks', in R. Forster and O. Ranum (eds), *Food and Drink in History*, Baltimore, Md., 1979. pp. 1–16.

3 D. Grigg, *The Dynamics of Agricultural Change: the Historical Experience*, London, 1982, p. 187.

4 D. Grigg, *The Transformation of Agriculture in the West*, Oxford, 1992.

5 W. L. Langer, 'Europe's initial population explosion', *American Historical Review*, **69**, 1963, pp. 1–17.

6 Arnold, *Famine*; Dando, *The Geography of Famine*; B. Ashton, K. Hill, A. Piazza and R. Zeitz, 'Famine in China, 1958–61', *Population and Development Review*, **10**, 1984, pp. 613–45.

7 *The Economist*, vol. 325, no. 7789, 1992.

8 World Bank, *The World Development Report 1992*, Oxford, 1992, p. 135.

Index